MANAGEMENT OF PROBLEM SOILS

Principles and Practices

MANAGEMENT OF PROBLEM SOILS
Principles and Practices

V. C. Srivastava
Ph.D., FISA
Ex- University Professor & Chairman
Department of Agronomy
and
Former Dean, Faculty of Agriculture
Birsa Agricultural University, Ranchi (India)

AGROBIOS (INDIA)

Published by:
AGROBIOS (INDIA)
Agro House, Behind Nasrani Cinema
Chopasani Road, Jodhpur 342 002
Phone: 91-0291-2642319, Fax: 2643993
E. mail: agrobios@sify.com

AGROBIOS (INDIA)

ISBN No. (13): 978-81-7754-432

Published by: Dr. Updesh Purohit for Agrobios (India), Jodhpur
Laser Typeset at: Yashee Computers and Printers, Jodhpur
Cover Design by: Reena
Printed at: Agrobios Digitals, Jodhpur

Dedicated to My Teachers

Prof. S. C. Mandal

Dr. K. K. Jha

Prof. B. N. Chatterjee

Prof. V. V. Rachinsky

Prof. A. D. Fokin

In appreciation of their inspiration and guidance

Preface

Concern about sustainability of land and environmental safety is increasing day by day. With the increasing population, it is also becoming clear that food security to teeming millions will not be possible unless the available resources are efficiently utilized for increasing the productivity. Soil is the most important resource for a sustainable agriculture, since it contains essential nutrients, stores the water and provides the medium in which plants grow. Soil as a filter, buffer and transformation agent between the atmosphere and the ground water strongly influences the plant nutrition, water cycle and gas exchange between terrestrial and life on earth. Problem soils present serious threat to our ability to increase food production to meet the expanding needs. Scientific management of soil-related constraints holds the key to sustainable food production and livelihood security on long terms basis.

This book on 'Management of Problem soils' makes an attempt to compile the available information in a cognet and easily understandable manner which has been a felt need of students, teachers, research workers and administrators in agriculture. The author has tried to discuss various aspects of management of problem soils from the point of view of the teachers explaining them to his class. It is hoped that the book will also serve as a good reference source for those interested in management of problem soils. The book contains chapters on important categories of problematic soils.

The author is deeply indebted to authors of books, research papers and bulletins on the subject from which most materials in the book has been drawn. The author welcomes suggestions from readers for improvement of this book in future editions.

Finally, my appreciation of the encouragement given by my wife 'Kiran' and sons Rajnish and Manish, the moving spirits responsible in this arduous and challenging task of writing the book is sincerely recorded.

V. C. Srivastava

Contents

CHAPTER 1

Introdcution

According to Norman H. Taylor, late director of the New Zealand Soil Bureau "the soil is a living body." It is finite, non-elastic and non renewable natural assets. Soil related factor are among the most significant environmental constraints on crop production responsible for the failure of crop varieties to live up to the expectation in post green revolution era. These factors have a major influence on the economy and distribution of farming population. The farmers through out the world have favoured areas with naturally highly productive soils and have shunned those with the less productive soils. Ignorance of improved technology and lack of inputs have constrained them from improving crop performance in areas with poor soils. The main reasons for unsustainable agricultural practices could be both cause and effect of soil degradation - a significant factor of environmental degradation.

Soil degradation means that the soil lose their natural fertility for the production of food and regenerative raw materials. It also means loss of soil capability to express its normal potential. At the same time substance cycles linking the soil with the vegetative cover, the surface water and ground water as well as with the atmosphere are severely affected and biological diversity is at risk. The most wide spread phenomenon is the loss of organic matter and the nutrients needed for a biomass production as a result of improper soil management together with wind and water erosion. There is increasing cause for concern about the deterioration of soil quality in the large parts of the earth due to salinization, acidification, waterlogging, increasing concentration of heavy metals and other persistent substances in the soil. Apart from restricting production, soil degradation promotes global warming. Green house effect is the outcome of smaller sink size due to sparse vegetation that these soils can effectively support.

The pace of soil degradation caused by human activity has accelerated all over the world in last few decades. Out of 1.5 billion ha of agricultural land currently used worldwide, roughly on-third has already been degraded as a result of overuse by human. The same applies to approximately one-fifth of 3.2 billion ha of grassland and 4 billion ha of woodland and savannah land.

In India, the demographic pressure is rapidly mounting on the natural resources. The unabated population growth pressure has negated the effect of green revolution of the 1970's. The large-scale degradation of natural resources has occurred due to population induced deforestation, conversion of forest land to agriculture, shifting cultivation, exploitation of fragile and marginal land to agriculture, intensive cropping, reduction in the frequency of fallowing, and indiscriminate use of chemicals. The present population of nearly one billion, accounting for about 18 per cent of world population, supported only 2.4 per cent of land area, is estimated to become 1.4 billion by 2025 AD and 1.7 billion by 2050 AD. It is estimated that India would need 325 million tonnes of food grains by 2025 AD as against 220 million tonnes at present to meet the requirement of burgeoning population and to achieve this target, production has to be increased by 7 million tonnes every year over the next 15 years without much increase in the net cropped area of 142.5 million ha (47% of geographical area) which is almost the upper limit to area extension for agriculture and there is no option except to vertical expansion or increasing productivity per unit area per unit time.

Out of India's total geographical area of 329 million ha, 187.8 million ha representing nearly 57 per cent presently suffer from various kinds of degradation, such as wind and water erosion, water logging, salinization, ravines, mines, quarrying and landslides. The total area of degraded soil comprises of 148.9 million ha affected by water erosion, 6.2 million ha by wind erosion, 3.7 million ha by chemical degradation, 10.1 million ha by salinisation, and 11.6 million by physical degradation. The large scale soil erosion washes away the top fertile soils and the newly exposed soil is relatively infertile with poor physical make up. Shifting cultivation is practised in the north-eastern hill regions on nearly one-third of cultivated area, resulting in loss of 18 million tonnes of soil. It is estimated that occurance of acid soils is more than 90 per cent in North-eastern states, 39.6 per cent in West Bengal and 28.2 per cent in Orissa. It addition to this acid soils are wide spread in other parts of India. Many of soils in humid regions are very acidic i.e. (pH<5) and crop growth is hindered by toxicities of aluminum or manganese or both. Crop production is drastically reduced when aluminum saturation of the active cation exchange site is greater than 60 per cent and tends to be optimum when aluminum saturation is zero. Similarly, India has about 23.2 million ha of saline soils, 2.5 million ha of alkali soils, 0.5 million ha of acid-sulphate soils, and 11.6 million ha of waterlogged soils. Accordingly the crop yields of these regions are not only low but also highly variable depending on the rain intensity, frequency of rainfall, soil type, topography and climatic conditions.

Effects of land degradation are far severe in semi-arid regions of India where over-grazed, drought prone ecosystem is fragile. Soils of arid and semi-arid regions remain bare during most part of the year. Severe constraints on bio-mass production is a characteristic common to rainfed agriculture. Accelerated wind erosion and desertification are largely a result of man made problems. Aridizations mediated wind erosion is closely linked with land misuse. Interventions like removal of vegetative cover, uncontrolled and excessive grazing, introduction of intensive agriculture and large unprotected field devoid of protective vegetative caps accelerate wind erosion.

The term 'management of problem soil' is used to cover many types of activities aimed at improving soil productivity or increasing crop production on specific problem soils. When a soil does not permit normal growth of crops under a given set of soil-water environmental conditions, it is called problem soil under those conditions. Suitable management practices are required to be developed to offset the barriers that are detrimental to a healthy crop growth. Problem soils include special types of soils with special problems, such as acidity, salinity, alkalinity, erosion or water logging. Hence, sustained crop production calls for proper understanding of nature and properties of problem soils and their management based on scientific research and approach to ensure the sustainability to the land and life on this planet.

CHAPTER 2

Acid Soils

Acid soils cover a very large area in the world. Podzols and other podzolic soils in temperate regions and laterites and other red soils in tropical and sub-tropical lands are acidic in reaction. Acid soils of India cover a very large fraction of country's land mass constituting about 30 per cent of cultivated land. Out of total geographical area of about 329 million hectares, the net cultivated area in India is nearly 149 million hectares and extent of soil exhibiting acidic reaction (pH < 7.0) may be around 90 million hectares (Mandal, 1995). An earlier estimate indicates that there are 49 million hectares of acid soils which is nearly one-third of cultivated area of India. Out of which 26 million hectare have pH below 5.6 and 23 million hectare between 5.6 and 6.5 (Raychoudhary, 1974). According to National Bureau of Soil Survey and Land Use Planning, Nagpur, out of 90m ha od acid soils in India strongly acidic (pH <4.5), moderately acidic (pH 4.5-5.5) and slightly acidic (pH 5.5 – 6.5) occupy 6.19, 24.81 and 58.94 m ha., respectively.

DISTRIBUTION OF ACID SOILS IN INDIA

Acid soils of India occur in greater Himalays, middle and lower Himalays, the eastern and north-eastern plains, penisular India and coastal plains under changing conditions of land-scapes geology, climate and vegetation. Acid soils are extensively found in region of high rainfall and temperature. North-eastern region of the country comprising Assam, Nagaland, Meghalaya, Arunachal Pradesh,

Mizoram, Tripura, Manipur, Sikkim, West Bengal (particularly Darjeeling and Jalpaiguri) has the largest stretch of acid soils.

In Jharkhand, about 70 per cent soils of Ranchi district, 60 per cent of Hazaribagh and Dhanbad district and 30 per cent of Palamu district are acidic.

Barring the coastal plains, the soils of other districts of Orissa are acidic and these soils are mostly concentrated in the upland district of Orissa. Acid soils are found in Purnea district of Bihar and narrow strip of *tarai* region. Soils in parts of Gurdaspur, Hoshiarpur and Patiala district of Punjab are acidic in nature. Acid soils are also found in Hilly and mountain regions of Uttar Pradesh and Jammu and Kashmir.

Rice growing soils of Chattisgarh (Bilaspur, Balaghat, Raipur, Durg and Bastar districts), Madhya Pradesh (Surguja, Shahold, Rajagarh, Mandla districts) and Maharashtra (Ratangiri, Kolhapur, Satara, Thana and Kolaba districts particularly) are acidic in nature and provide positive correlation with high temperature and rainfall.

In south India, acid soils are found in part of Tamilnadu, Andhra Pradesh, Mysore and Shimoga districts of Karnataka. Apart from rainfall and temperature, rich organic deposits in the coastal state of Kerala have resulted in the formation of peats under marshy conditions responsible for acid soils in the state. Acid sulphate soils of Kuttanad (Kerala) and Sunderbans (West Bengal) are influenced by dissolved chemicals, such as H_2SO_4, $Al_2(SO_4)_3$ and $Fe_2(SO_4)_3$ and contribute to pH dependent acidity.

DEVELOPMENT OF ACID SOILS

The major processes responsible for development of acid soils are (Mandal *et.al.*, 1975):

a) laterization of varying degree,
b) podzolization in areas with sub-temperate to temperate climate,
c) intense leaching in light textured soils, and

d) marshy conditions with significant amount of partial decomposed organic matter.

Although origin of soil from acid parent materials, such as granite, are important for development of acid soils, the factors which are particularly dominant in the development of acid soils are rainfall, temperature, hydrological conditions and native vegetation which determine the degree of soil acidity. The existing differences in climate, vegetation, parent materials, topography and the degree of interaction among different soil forming factors induce a wide range of variation in minerological, physical and biological properties of acid soils.

In a humid climate, acidic rocks produce unsaturated clay colloids and ultimately acid soils. In case of basic rocks, weathering of minerals are mainly by hydrolysis, and base hydroxides, soluble silica and clay minerals are the products of weathering. In the presence of abundance of basic hydroxides, the clay minerals tend to be base saturated. But as the weathering of minerals cannot keep pace with loss in leaching, the clay minerals cannot be completely base saturated. When the major portion of primary mineral is weathered, the supply of base hydroxides are far less than what is lost in leaching. In this case the H^+ ions of the dissociated rain water replace the adsorbed basic cations from the clay complex and the soil becomes acidic.

The Sedentary soils of Jharkhand and alluvial soils of north-eastern part of Purnea district of Bihar exhibit acidic reaction. The sedentary soils have developed on rocks of very ancient origin, namely - Archeans, Gondwanas, Vindhyans and their various subsystems. These rocks have weathered to give parent materials for soil formation of varying minerological and chemical composition. Influence of other factors of soil formation, such as climate, vegetation etc. have acted on these parent materials situated on the land surfaces of extremely varying topography over different length of time to form soils of distinctly different characteristics. The acidic soils of Purnea district are developed in the flood plains of Mahananda and Kosi rivers. These rivers being short in length have deposited sandy alluvium in the area. The continuous and intense leaching of the sandy soils due to very heavy rains (1250 to 2500 mm) have caused acidity in these soils.

A base saturated soil can form acid soils under better drained condition by removal of bases and replacement by H^+ ions under the following conditions:

i) **Change in climate:** If the climate of an area becomes semi-humid to humid from semi-arid to arid conditions, the dissociated H^+ ions of the rain water will remove and replace the adsorbed exchangeable bases from the clay mineral surface and the soil will become acidic due to abundance of adsorbed H^+ ions

ii) **Introduction of irrigation:** When excessive irrigation water is applied which may be lost in drainage the result will be the same as in case of change in climate from arid to humid.

iii) **Introduction of drainage in waterlogged areas in humid climate:** Along with the drained water the adsorbed bases which are replaced by the H^+ ions of the rain water and carbonic acid are also lost. Thus, on losing adsorbed bases and their replacement by hydrogen ions the soils turn acidic.

iv) **Application of acidic and acid forming fertilizers in soils having no or very low reserve of $CaCO_3$:** Fertilizers like ammonimum sulphate, ammonium chloride, ammonium nitrate, urea and ammonium phosphate, etc., which have not metallic base produce acids either by dissociation or by oxidation of ammonia to nitric acid or both. The H^+ ions, thus produced, replace the bases present on the clay colloids and turn the soil to acidic. The adsorbed NH_4 ion is oxidised into HNO_3 by nitrifying bacteria HNO_3 dissociates into H^+ ions and NO_3 ions of which the H^+ replaces bases on the clay surface and the $Ca(NO_3)_2$ thus formed is either taken up by the plants or lost in leaching.

In the case of organic soils, acidity develop under better drainage conditions. The content of H^+ ions of water or its ionization increases as it dissolves CO_2 of the atmosphere or soil, in fact it becomes weak solution of carbonic acid. Organic matter also impacts acidity by the dissociation of carboxylic, amino acid and other acidic radicals present on them. Organic matter also produce organic acids like citric, oxalic and acetic acids as the intermediate products in its auxerolic decomposition, which impart acidity in peat, marshy and sandy soils.

Cat clays or mineral soils, where acidity is due to the presence of free H_2SO_4 are produced by the drainage of sulphide rich swamps and marshes. These metallic sulphides are oxidised to H_2SO_4 by sulphur oxidising bacteria *Thiobacillus thioxidaus.*

Microbial activity responsible for many processes, such as decomposition of organic residues and nitrification have been known to increase soil acidity.

CLASSIFICATION OF ACID SOILS

Broadly the acid soils of India can be classified into seven distinct groups, namely (1) laterite (pH 4.8 - 7.0), (2) laterite and lateritic red (pH 5.0 - 7.0), (3) mixed red and black / yellow (pH 5.5 - 6.5), (4) ferruginous red (pH 5.0 - 6.8), (5) Podsolic, brown forest and forest (pH 4.5 - 6.5), (6) foothill soil analogous to Wiessenboden (pH 4.5 - 6.5), and (7) peat soils (pH 3.5 - 4.5).

HYDROGEN ION CONCENTRATION AND pH

Water is dissociated into hydrogen ion and hydroxyl ions:

$$H_2O \rightleftharpoons H^+ + OH^-$$

Freshly distilled water contains the same number of H^+ and OH^- ions. As such, it is neutral in reaction. When the soil solution contains the same number of H^+ and OH^- ions it is also neutral. When an acid is dissolved in water it will dissociate and release hydrogen ions (H^+), which makes the solution acid. Similarly, when an alkali is dissolved in water, it produces hydroxyl ions (OH^-), which makes the solution alkaline. Although equivalent quantities of all acids contain the same amount of total hydrogen, when dissolved in water, the hydrogen in them does not ionise to the same extent. Acids are called weak or strong depending upon the degree of ionisation. The greater the ionisation, the larger the number of hydrogen ions in solution and the stronger the acid. Hydrochloric acid is a strong acid while acetic acid is a weak acid, even though the titrable acidity (total acidity) in the two acids may be the same.

The most convenient method of expressing the relationship between H^+ and OH^- is pH. pH is defined as the logarithm of the reciprocal of the hydrogen-ion concentration in grams per litre usually written as:

$$pH = \log \frac{1}{(H^+)}$$

At neutrality, the hydrogen-ion concentration has been determined to be 0.0000001 or 1×10^{-7} grams of hydrogen per litre of solution. Substituting this concentration into the formula, we get

$$\begin{aligned} pH &= \log \frac{1}{0.0000001} \\ &= \log 10{,}000{,}000 \\ &= 7 \end{aligned}$$

At a pH of 6, there would be 0.000001 grams of active hydrogen or 10 times the concentration of H^+ compared to pH of 7 (0.000001 g of H per litre). At each smaller pH unit, the (H^+) increases by 10 in concentration. It, therefore, follows that a pH of 6 is 10 times more acidic than a pH of 7, a pH of 5 is 10 times more acidic than a pH of 6, and so on.

The pH concept permits the expression of the hydrogen ion concentration on a scale of acidity and alkalinity from 0 to 14. pH 7 is neutral, lower values like pH 6 or less indicate acidity, and higher values like pH 8 or more show alkalinity.

(I) SOIL ACIDITY

Concept

Soil acidity is a base unsaturated soil which has got enough of adsorbed exchangeable H^+ ions and probably Al^{+3} ions so that to give it a pH of lower than 7. In certain cases this acidity may also exists in form of free organic acids especially in soils rich in organic matter (peat soils) and organic matter rich sandy soils. Likewise in certain soils (called cat clays by Dutch Scientists) the acidity is due to the presence of free H_2SO_4 (Sulphuric acid), a product of sulphide oxidation.

The activities of various forms of soil acidity are permanent charge H^+, permanent charge Al^{3+}, hydroxy-Al-monomers, hydroxy-Al-polymers, organic matter acidity, lattice Al-OH and Si-OH. Soil acidity should be viewed in terms of specific proton donors viz., (i) monomeric-alumino - hydronium (ii) polymeric positive edge, and (iii) functional group of organic matter, etc. (Mandal, *et.al.,* 1975). Hence, soil acidity many be defined as soil system's proton yielding capacity and its neutralization should be considered according to its proton retention sites as well as reactions leading to donation of proton (Barthakur, 1990).

Division into Al^{3+} and H^+ (exchange acidity) can be based empirically upon soil pH. Exchangeable H^+ is normally present in measurable quantities in the soil only at pH value less than 4.0. Hence, exchangeable H^+ is of concern only for extremely acidic soils. According to Marion *et al.,* (1976), in an acidic medium the Al^{3+} ion is dominant below pH 4.7, $Al(OH)_2^+$ between pH 4.7 and 6.5, Al $(OH)^0_3$ between pH 6.5 and 8.0 and Al $(OH)^-_4$ above pH 8.0. In the pH range of 5.5 to 7.0 hydroxy-Al-polymers predominate among acid soil components, exchangeable acidity is virtually absent and only titrable acidity are present in the measurable quantities. Bhan *et.al.,* (1979) reported that significant quantities of weak acidic R-COOH and R-OH groups on soil organic matter and incompletely neutralized hydroxy-Al-polymers can be present in soils of pH more than 7.0.

Theoretically any soil having a pH lower than 7 should be called an acid soil, but for practical purposes as related to plant growth the pH ranges for a neutral soil and different degree of acidity has been stated below:

Soil reaction	**pH range**
Neutral	6.6 - 7.3
Slightly acid	6.1 - 6.5
Medium acid	5.6 - 6.0
Strong acid	5.1 - 5.5
Very strong acid	4.6 - 5.0
Extremely acid	4.5 to below

(Adopted from USDA Year book of Agriculture, 'Soils', 1957)

Constraints to Crop Production

The major constraints to crop production in acid soils are toxicities of aluminium, iron, maganese and hydrogen and deficiencies of calcium (Jackson, 1967). Toxic concentration of iron in acid soils causes purplish and reddish brown mottling (bronzing) in some cultivars and oranging or yellowish spreading downwards from the tip to the older leaves followed by dying of the leaves. Roots are scanty, coarse and often dark brown from the coating of ferric oxide (Ponnamperuma *et al.,* 1955). The symptom of aluminium toxicity in rice has been described as an interveinal orange yellow discolouration of the tips of the oldest leaves followed by brownish mottling. (Tanaka and Navasero, 1966).

According to Kamparath (1980), crop production is drastically reduced when aluminium saturation of the active cation exchange sites is greater than 60 per cent and tends to be optimum when aluminium saturation is zero. When a soil is more acidic than 6.0 to 6.5 pH, the availability of plant nitrogen, phosphorous, sulphur, calcium, potassium, magnesium and molybednum may be too low for satisfactory plant growth and development. Such soils are generally characterised by light textured surface with high permeability and low water holding capacity. The beneficial activity of Azotobactor and nodule forming bacteria of legumes are adversely affected.

Soil reaction preferences of field crops

Few plants grow well in wide range of pH, while others do well only within narrow limits. A large number of field crops have their optimum ranges around pH 6.5. The approximate optimum pH ranges of the major field crops are given in Table 1. Lucerne, sweet clover, cabbage and cauliflower thrive under slightly acidic to alkaline conditions. Potatoes grow well in strongly acidic to neutral soil. Some of the grasses, tea, coffee and fruit tress (blue berries and cranberries) thrive well on strongly acidic soils.

TABLE 1: Soil reaction preferences of field crops

	Crops	Optimum pH range
A.	**Cereals**	
	Rice (*Oryza sativa*)	5.0 - 6.5
	Wheat (*Triticum sp.*)	5.5 - 7.5
	Maize (*Zea mays*)	5.5 - 7.5
	Millets	5.0 - 6.5
	Oats (*Avena sativa*)	5.0 - 7.5
	Barley (*Hordeum sativum*)	6.5 - 8.0
	Sorghum (*Sorghum bicolor*)	5.5 - 7.5
	Rye (*Secale cereale*)	5.0 - 7.0
B.	**Pulses and legumes**	
	Lima beans (*Phaseolus vulgaris*)	6.0 - 7.0
	Field beans (*Phaseolus limensis*)	6.0 - 7.5
	Berseem (*Trifolium alexandrinum*)	6.0 - 7.5
	Lucerne (*Medicago sativa*)	6.2 - 7.8
	Peas (*Pisum sativum*)	6.0 - 7.5
	Soybean (*Glycine max*)	6.0 - 7.0
	Cowpeas (*Vigna unguiculata*)	5.0 - 6.0
	Groundnut (*Arachis hypogaea*)	5.3 - 6.6

	Crops	Optimum pH range
	Chickpea (*Cicer arietinum*)	6.0 - 7.5
	Pigenpea (*Cajanas cajan*)	5.5 - 7.0
	Lentil (*Lens culinaris*)	5.5 - 7.0
C.	**Miscellaneous Crops**	
	Sugarbeet (*Beta vulgaris*)	6.5 - 8.0
	Cotton (*Gossypium hirsutam*)	6.0 - 7.5
	Linseed (*Linum usitatissimum*)	5.0 - 7.0
	Potatoes (*Solanum tuberosum*)	4.8 - 6.5
	Sunflower (*Helianthus annuus*)	6.0 - 7.5
	Tobacco (*Nicotiana sp.*)	5.5 - 7.3
	Sugarcane (*Saccharum officinaram*)	6.0 - 8. 0
	Tea (*Thea sinensis*)	4.0 - 6.0

CHARACTERISTICS OF ACID SOILS

These soils are mostly base unsaturated, low-water retentive and nutritionally imbalanced. Soils with high permeability and low clay activity are improved by passing heavy rollers affecting compaction. Crust formation is managed by mulching.

Physical Characteristics

Acid soils which have very little amount of electrolytes are in slightly dispersed condition and as such clay may be bodily carried downwards in the profile and form pan. These soils also do not form as good water stable aggregates as in the case of neutral soils. Heavy acid soils may also get puddled.

Acid red loams of Jharkhand indicate that (i) the surface soils are generally light in texture, have high permeability and low water holding capacity. (ii) The leaching of clay of lower horizons is apparent. Cultivation of unterraced land results in the loss of clay and other colloidal matter (iii) These soils are well aggregated mainly due to the presence of hydrated oxides of iron and almunium. Growing of grasses improve water stable aggregates of these soils. Acid soils of eastern Purnea district of Bihar are also of very light texture with high permeability, low water holding capacity but poor aggregation (Mandal and Sinha, 1963).

The soils of Bramaputra valley in Assam is classed as sandy loams. Old alluvial hill soils are sandy loam to clayey in texture, while new alluvial soils are mostly sandy loam, silt loam or clay loam. The acid soils of Chattisgarha are light sand soils. Occasionally medium and heavy textured soils are also encountered (Raychaudhary *et. al.,* 1963). Acid soils of Madhya Pradesh and Punjab ranges from sandy to clay loam. Studying the effect of compaction of laterite soils of Kharagpur (West Bengal) of sandy loam texture, Varade and Ghildyal (1967) indicated increased bulk density and hardness with a simultaneous decrease in hydraulic conductivity and non-capillary porosity. Yadav and Banerjee (1968) reported that in acid soils of forest area of Dehra Dun (Uttranchal), the amount of total water stable aggregates in the surface soil was found to decrease gradually with the depth of profile.

Chemical Characteristics

1) **Base Unsaturation**: Acid soils have lesser amounts of basic cations than H^+ ions adsorbed on the clay colloids and thus they are base unsaturated. More acid the soils, greater is the base unsaturation or the amount of exchangeable H^+ ions and lower the pH. But different kinds of soil give different pH at the same percentage base saturation, which is dependent on the nature of clay minerals present in the soil. Soils having montmorillonite clay give a lower pH than having kaolinitic clay at the same amount of base saturation.
2) **Buffering capacity:** In a soil suspension, H^+ ions are present as exchangeable hydrogen on the clay surfaces and only a small amount is present in the soil solution. When the pH of soil suspension is measured, it is only the concentration of H^+ ions in soil solution is measured. As soon as the H^+ ions of soil solution is tried to neutralise by OH^- ions of alkali, an equivalent amount adsorbed H^+ ions are released from the clay colloids into the soil

solution. Thus, it prevents the rise of pH of the soil solution equivalent to the amount of alkali added. This happens till all the adsorbed H^+ ions on the clay surface is released and neutralised. This phenomenon is called the **buffering capacity** of a soil, and the acidity due to H^+ ions in soil solution is called **active acidity**. The acidity which is produced by the release of adsorbed H^+ ions is called **potential or reserved acidity**. This potential or reserve acidity is the cause of buffering in an acid soils. The buffering capacity or reserve acidity of a soil depends on the nature of clay minerals present in a soil. The soil having montmorillonite clay minerals has more buffering capacity than those having illite or kaolinite clay minerals in them.

3) **Causes of poor plant growth**: As the acidity of soil increases many plants grow poorly and they do not grow at all in an extremely acid soils. This may be due to the presence of toxic substances or low availability of plant nutrients or the combined effect of both in an acid soil.

   a) Presence of toxic substances:

      i) High concentration of H^+ ions in acid soils are detrimental to plant growth because they hinder the uptake of other cations.

      ii) Soluble Al ions occuring in acid soils exert toxic effect on most of the plant species. Soluble Al restricts or stops root growth at solution concentration as low as 1 ppm.

      iii) In acid soils manganese occur in divalent or soluble form. With the increase in acidity the solubility of manganese increases. Different crops tolerate different concentrations of Mn. However, under reduced conditions prevailing in flooded and partially inundated acid soils can result in accumulation of soluble Mn to a concentration as high as 100 ppm, which may be toxic for plant growth.

      iv) Iron is less soluble than Al and Mn and hence, there are less chances that they are in toxic concentration in the acid soils used for agriculture. Iron is highly soluble below pH 2.5, a soil with such a low pH has not been observed in natural condition.

4) **Availability of plant nutrients**: The degree of soil acidity, depletion of exchangeable bases and availability of plant nutrients varies widely in acid soils depending on climatic conditions and land use practices. Soil erosion in high rainfall regions further enhances the depletion of plant nutrients from acid soils. Effects of pH on plant nutrient availability in soils are complicated and inter-related. However, some generalization can be made.

   a) **Nitrogen**: The organic matter and nitrogen content of acid soils show extremes of variation depending upon the nature of vegetation and intensity of organic matter decomposition. In general, laterites, lateritic and red loam soils are characterised by low status of organic matter and nitrogen, however, hill and forest soils are comparatively richer in organic matter and nitrogen. Nitrogen availability in acid soils is the pH dependence of ammonium ion fixation between the lattices of expanding layer silicate minerals. Such fixation generally decreases with decreasing pH.

   b) **Phosphorous**: Most of the acid soils in India are poor in phosphorus. In acid soils, where soluble aluminium and iron are higher than in neutral soils, phosphorous combines with soluble almunium or iron and forms relatively insoluble almunium and iron phosphates, which on aging and drying turn into very insoluble form. Thus, the deficiency of phosphorous due to its fixation may be real cause for low plant growth. Plants with phosphorous deficiency generally have phosphorous content below 0.1 percent.

   c) **Potassium**: Most of the acid soils are moderate to high contents of potassium as in acid medium it is released both from the clay minerals and primary potash bearing minerals. However, in sandy soils low in potash bearing minerals, a deficiency of potassium is likely to occur in acid soils. Whatever potassium is released in these soils are likely to be leached away with rains. Potassium deficiency occurs in the sandy acid soils of north-eastern Purnea district of Bihar. Mandal *et. al.*, (1975) found that acid soil of sedimentary origin containing abundance of mica and K-bearing minerals are relatively richer in available potassium content.

d) **Micro-nutrients**: The availability of micro-nutrients, like Cu, B, Fe, Zn and Mn increases with increasing acidity. However, they may be deficient in sandy acid soils due to leaching. The availability of Mo decreases as the pH decreases. Acid soils of the north-eastern region of the country are deficient in B and Mo (Panda, 1987).

Microbiological Characteristics

1) **Microbial population**: In acid soils with decrease in available phosphorus and increase in H^+ ions, the nature of microbial population also differs. As soil becomes more acid there are increase in number of fungi over bacteria. In strongly acid soils the dominant form of microorganism in fungi.
2) **Decomposition of organic matter**: With increasing acidity there is decrease in the rate of organic matter mineralization, which increases with application of lime, phosphorus or both.
3) **Nitrification**: Nitrification (i.e. oxidation of NH_4^+ to NO_2^- to NO_3^-) occurs rapidly in neutral soils and decreases with increase in acidity. The rate of nitrification depends on supply of calcium and phosphorus and these are in short supply in acid soils. In organic soils, nitrification occurs even at a low pH of 3.8 as these soils contain enough phosphorous and exchangeable calcium for the microorganisms.
4) **Nitrogen fixation**:
   a) **Symbiotic N-fixation**: Most of legumes which have nodules in their roots prefer neutral or slightly acidic soils and are not able to grow properly in soils having low pH. Though nitrogen fixation by *Rhizobium species* depends upon the growth habits of the associated crops, very little symbiotic

nitrogen is fixed in acid soils with low pH, as condition is not favourable to the growth of their host plants.

b) **Non-Symbiotic N fixation**: Except *Azotobactor indicum* all the other species of Azotobactor rarely exist below pH 6. They are very sensitive to phosphate deficiency which is always in short supply in acid soils of low pH. Hence, there is very little nitrogen fixation by these species.

5) **Plant diseases**: The predominant microorganisms is acid soils with low pH are fungi. As such, diseases that are caused by fungi are more common in such soils, whereas, occurance of diseases that are caused by bacteria and actinomycetes are less.

AMELIORATION OF ACID SOILS

Liming

Although growing of acid tolerant plant species and varieties is suggested for acid soils, liming is recognised as a standard management practices for increasing the productivity of such soils.

Liming means adding to the soil any calcium or calcium and magnesium containing compound that is capable of reducing the acidity of soil. Lime correctly refers only to calcium oxide CaO, but the term as applied to agriculture is universally used to include various other materials also, like calcium carbonate, calcium hydroxide, calcium magnesium carbonate and calcium silicate slags.

When lime is added to moist soil, the soil solution becomes charged with calcium ions. These active Ca ions exchange places with hydrogen ions in the exchange complex. Hydrogen combines with OH ions to form neutral water, or with CO_3 or HCO_3 to form unstable H_2CO_3, which is readily changed to H_2O and CO_2.

Lime is added to acid soils primarily for three purposes:

i) To supply calcium and magnesium as a plant nutrient.
ii) To reduce the toxicity of aluminium, maganese and iron.

iii) To increase the pH of acid soils and thereby make other plant nutrients more available.

Benefits of liming: The benefits of liming on acid soils are enlisted below:

i) It neutralises H^+ ions and the increase of OH^- ions neutralises soil acidity by raising soil pH.
ii) It reduces the base-unsaturation of the acid soils, whereby Ca and Mg ions are adsorbed on the clay colloids replacing H and Al ions. Thus Ca, Mg and other bases become more available to the plants.
iii) It decreases the solubility of aluminum, magnese and iron ions, and thus reduces their toxic effect on plants.
iv) It renders phosphorous and molybdenum in more available forms. Phosphate fixation by soluble Al and Fe is stopped. Phosphate is a major nutrient required by plants. It facilitates symbiotic nitrogen fixation. Mo apart from being micro nutrient is a must for N-fixing organisms.
v) It decreases the solubility of Zn, Cu, and B in such acid soils where they may be in toxic concentration.
vi) Some potassium may also be released for plant utilization.
vii) It increases the rate of decomposition of organic matter by which more plant nutrients are released.
viii) Nitrification of NH_4^+ to NO_3 is increased.
ix) It favours the growth of leguminous plants, whereby more symbiotic nitrogen is fixed in the soil.
x) Plant diseases, which are favoured in acid conditions are either checked or considerably reduced.
xi) It favours the formation of crumb structure and thereby increases aeration, root penetration and water holding capacity of soils. It improves the flocculating power of soil particles, whereby more water stable aggregates are produced.
xii) The effectiveness of fertilizers and manures is increased. Ammonical nitrogenous fertilizers are readily converted to

nitrates and fixation of phosphate fertilizers by soluble Al and Fe is checked.

xiii) It helps in preventing the destruction of clay minerals taking place in acid soils.

Harmful effects of overliming

In light soils, where there is low amount of clay minerals and as such low buffering capacity, the harmful effects of overliming are immediately seen. Likewise in tropical soils (laterites and latosols) which have predominance of sesquioxides over silicate clay minerals and low organic matters, the buffering capacity is low and the overliming effect may set in a bad form. The harmful effects of overliming are given below:

i) Reduces the supply of Fe, Mn, B, Zn and Cu which are fixed in someway or other causing deficiency.
ii) Phosphates may get converted into different kind of apetite and thus, become again unavailable to the plants or availability may be reduced to a critically low level.
iii) Where liming material is nearly pure $CaCO_3$, magnesium deficiency may also set in.
iv) Potassium uptake may be reduced.

The ultimate result may be a reduction in yield and not the real benefit which could have been got by judicious liming.

Overliming injury may be reduced by application of large amounts of manure, green-manure crops, compost, phosphorus fertilizer, boron or a mixture of micro-nutrients.

Crop response to liming

The direct and indirect effects of liming generally lead to increased crop production, the magnitude of which varies widely under different soil conditions and cropping pattern. The beneficial effects of liming of acid soils may be attributed to amelioration of toxic effects of exchangeable Al, Fe and Mn, enhanced availability of plant nutrients and increased efficiency of applied fertilizers.

On the basis of the degree of response to lime, the crops have been grouped into the following three classes (Mandal *et al.*, 1966):

i) **High response class**: Pigeonpea, soybean and cotton.
ii) **Medium response class**: Chickpea, lentil, pea, groundnut, maize, sorghum, and wheat.
iii) **Low response class**: Barley, finger millet, little millet, rice, mustard, etc.

Although rice is highly tolerant to soil acidity, it responds to liming in strong acidic soils.

The yields of cereals, pulses, cotton, groundnut, fodder crops and vegetable showed an increase to the extent of 5 to 16 folds by liming acid soils at Ranchi. Response to fertilizers also increased considerably by liming acid soils. However, upland rice, potatoes, small millets, etc. did not show any response to liming (Mandal and Jha, 1970).

The existence of differential response of crops to liming in different parts of the country suggests that the cropping pattern may be adjusted in limed soils to realise the maximum benefits from the investment on liming. It is known that the effect of lime applied in full dose lasts for 4 to 5 years which decline gradually in the last two years. Hence, under the rainfed conditions, crops like groundnut, soybean, cotton, pigeonpea, etc., are highly responsive to liming (pH 6.5-7.0) and as such may be grown in the first two years of lime application followed by medium responsive (pH 6.0 - 6.4) crops, such as maize, sorghum, blackgram, greengram, etc. In the 5th and 6th yeas when the effect of lime is greatly reduced (pH below 6.0) low responsive crops, like rice, small millets, potato, etc., may be grown (Mandal, 1962, 1971). They have also indicated the promise of short-duration varieties in monsoon season followed by drought resistant crops, e.g., linseed, mustard, ratoon sorghum, chickpea, lintil, barley, etc. in winter under rainfed conditions preferably with life-saving irrigation making use of water stored during the rains.

Lime requirements

The principal factors which determine the lime requirement of acid soils are

i) pH
ii) amount and type of clay
iii) cation exchange capacity, and
iv) buffering capacity

The change in soil pH is dependent on its buffering capacity and the buffering capacity of different soils differ widely depending on the amount and nature of clay minerals and organic matters. Hence, to raise the pH of two soils, say from 5 to 7, having different amount of clay or organic matter or having the different qualities of the same will require different amount of the same liming material. To calculate lime requirement of acid soil various methods both direct or indirect are used as follows:

i) **Indirect method based on pH and texture of soils**: This is only an approximate estimation and is not expected to give accurate results. The lime requirement increases with increase in the clay content or fineness of soil.
Soils with less than average amount of organic matter will require less lime, and soils with more than average amount of organic matter will require more lime than the amount indicated in the table 2.

TABLE 2: Limestone needed to increase the pH of each soil textural class by one unit

Soil Texture	Limestone per Acre (one ha = 2.471 acre) Required to Increase the pH value by One whole unit (Tons)
Sand	1.0
Sandy Loam	1.5
Loam	2.0
Silt Loam	2.5
Clay Loam	3.0
Clay	3.5

The quantity of lime indicated is only for initial application. After three years the soil should again be tested for adding required amount of lime.

ii) **By actual and direct titration and formation of titration curves**: Moist incubation is done with graded amount of $CaCO_3$ or $Ca(OH)_2$ and the amount liming material required is determined to bring the soil reaction to a reference state from the potentiometric titration curve so obtained.

In actual practice in the field more lime is required than calculated from laboratory test to raise the pH to the same extent because the lime added to the filed is not limited to 6" or 15 cm as calculated in the laboratory and some is lost to drainage. Hence, the laboratory figure is multiplied by a figure called **lime potential** which shows the discrepancies between the field and laboratory data. The lime potential varies between 1.4 to 3. For most of the mineral soils it is between 1.6 to 2.0.

iii) **By using modified pH meters (also called lime meters):** It can directly read the amount of lime required for an area. Special buffers are used which react with all the exchangeable soil acidity (both reserve and active) and the drop in pH is measured from which lime requirement is calculated. These methods are based on schofield's lime potential formula:

$$pH - \frac{1}{2} p (Ca + Mg) = K$$

iv) **Rapid lime requirement methods** involving equilibrium extraction of soil with strongly buffered system followed by determination of the exchange acidity in the extract by direct titration or by measuring the change in pH of the extract. But strongly buffered solutions are unable to indicate appreciable change in pH of weakly buffered soils high in extractable Al. Kemprath (1970) has suggested to base lime rates by relationship Al/100 g x 1.5 m.e. $CacO_3$/100 g which will give an exchangeable Al saturation of 15 per cent or less. With crops, very sensitive to Al, factor 2 should be used to determine the amount of lime required.

Apart from above mentioned methods the various other methods of determination of lime requirement are: (i) Mehlich's $Bacl_2$ - triethanolamine (pH 8.2) Method (1939), (ii) Woodruff's Ca-acetate-p-nitrophenol Method (1948), (iii) Riera's moist incubation with $CaCO_3$ (1946), (iv) Woodruff's Ca-acetate-p-nitrophenol Method (1948), (v) Pratt and Blair (p-nitrophenol-Ca-acetate + TEA) Method (1961), (vi) Rich Ca-acetate-triethanolamine Method (1955), (vii) Hutchinson and Mc Lean's Method, (viii) John's Method, (ix) Puri's method, (x) Brown's procedure and others.

Laboratory incubation studies indicated that about 450 Kg of lime per ha is required for increasing the pH by 0.1 unit in acid soils of Chotanagpur (Jharkhand) of similar texture. An additional 450 Kg. of lime per ha is required for meeting the requirement of the soils of next heavier textural class at the same pH level. A tentative schedule of lime dose according to soil pH and texture was worked out at Agricultural Research Institute, Kanke, Ranchi (Mandal and Jha, 1965) which is given in table 3 below:

TABLE 3: Lime requirement of soils of different texture (q/ha)

pH	Sandy loam	Loam	Clay loam
5.0	55.5	60.0	64.5
5.2	46.5	51.0	55.5
5.4	37.5	42.0	46.5
5.6	28.5	33.0	37.5
5.8	19.5	24.0	28.5
6.0	10.5	15.0	19.5
6.2	6.5	8.5	10.5

Liming material

More than 90 per cent of agricultural lime is calcium carbonate, some is calcium and magnesium carbonate, and a much smaller quantity is calcium oxide, calcium hydroxide or wood ash. The different liming materials are used wherever easily available and cheaper in cost. The chemical relationship of three important liming material is shown in the equation:

$$\underset{\text{(Limestone)}}{CaCO_3} \xrightarrow{-CO_2} \underset{\text{(Burnt lime)}}{CaO} \xrightarrow{+H_2O} \underset{\text{(Hydrated lime)}}{Ca(OH)_2}$$

The common liming materials are:

i) Calcic limestone ($CaCO_3$), which is ground limestone.
ii) Dolomitic limestone [$CaMg(CO_3)_2$], from ground limestone high in magnesium.
iii) Quicklime (CaO), which is burnt limestone, also known as burnt lime, oxide lime, caustic lime or stone lime.
iv) Hydrated (slaked) lime [$Ca(OH)_2$], coming from quicklime which has changed to the hydroxide form on reaction with water.
v) Chalk ($CaCO_3$), resulting from soft limestone.
vi) Coral shell lime is usually recommended as a liming material in Ceylon (Sri Lanka) due to its availability in abundance.
vii) Blast furnace slag ($CaSiO_3$ and Ca_2SiO_4), a byproduct of the iron industry. Some slag contains phosphorous and a mixture of CaO and $Ca(OH)_2$. The product is called basic slag and valued for its phosphorous content. It contains 6-8 per cent silicate, 5-7 per cent MgO and 43 to 48 per cent CaO.
viii) Miscellaneous sources, such as wood ash, ground oyster shell, press mud and by-product lime resulting from paper mills, sugar factories, tanneries, and water-softening plants and by-product $CaCO_3$ from fertilizer factories using gupsum process have potentiality as amendments for acid soils.

Calcitic and dolomitic limestone though have proven efficiency, having industrial use are not economically viable as amendment. But stromatolitic limestone with high silca contents is a reasonable agricultural lime containing 0.5 per cent phosphorous. Rock phosphate with higher CO_3^{2-} / PO_4^{3-} molar ratio, F/P_2O_5 weight ratio and absolute citrate solubility (ACS) can be applied directly. Partially acidulated rock phosphate and suitable mixture of water soluble P source and citrate insoluble P source (rock phosphate) may be applied to satisfy the adsorptive capacity of the acid soils economically (Panda, 2000).

All the liming materials mentioned have value for supplying calcium or calcium and magnesium, raising the pH, and making aluminium, mangnese, and iron less toxic.

Correction of Soil Acidity

i) **Ground limestone**: When it is applied to an acid soil, a part of calcium carbonate undergoes solution and combines with carbonic acid to form calcium bicarbonte.

$$\underset{\text{(Limestone)}}{CaCO_3} + \underset{\text{(Carbonic acid)}}{H_2CO_3} \longrightarrow \underset{\text{(Calcium biocarbonate in solution)}}{Ca(HCO_3)_2}$$

Calcium bicarbonate in solution form reacts with soil colloids, freely evolving carbon dioxide.

$$\underset{\text{(Acid Soil)}}{\begin{matrix} H^+ \\ H^+ \end{matrix} \boxed{\text{Micelle}}} + Ca(HCO_3)_2 \longrightarrow \underset{\text{(Normal Soil)}}{Ca^{++} \boxed{\text{Micelle}}} + 2H_2O + 2CO_2$$

While some portion of the limestone goes into solution, a part of the applied limestone comes in close contact with the soil colloids in solid condition. Under these circumstances, the adsorption of calcium with soil colloids also occurs.

$$\underset{\text{(Acid Soil)}}{\begin{matrix} H^+ \\ H^+ \end{matrix}\boxed{\text{Micelle}}} + \underset{\text{(Solid phase)}}{Ca(HCO_3)_2} \longrightarrow \underset{\text{(Normal Soil)}}{Ca^{++}\boxed{\text{Micelle}}} + H_2O + CO_2$$

As the above reactions of limestone proceeds, carbon dioxide is freely evolved, and the adsorption of the Ca and Mg raises the percentage base saturation of colloidal complex and thus, pushing up the pH of the soil solution.

ii) **Quicklime (burnt lime)**: When limestone is heated to a temperature of over 800°C, the calcium carbonate decomposes, yielding calcium oxide.

For 100 kg of calcium carbonate burned, 56 kg of calcium oxide is formed.

When quicklime (CaO) is added to any acid soil, it changes to bicarbonate form as it goes into solution.

$$\underset{\text{(Quick Lime)}}{CaO} + \underset{\text{(Water)}}{H_2O} \longrightarrow \underset{\text{(Hydroxide of lime)}}{Ca(OH)_2}$$

$$\underset{\text{(Hydroxide of lime)}}{Ca(OH)_2} + \underset{\text{(Carbonic acid)}}{2H_2CO_3} \longrightarrow \underset{\text{(Calcium biocarbonate)}}{Ca(HCO_3)_2} + \underset{\text{(Water)}}{2H_2O}$$

Calcium bicarbonate thus formed reacts with the soil colloids and calcium ions are adsorbed in the colloidal complex.

$$\underset{\text{(Acid Soil)}}{\begin{matrix} H^+ \\ H^+ \end{matrix}\boxed{\text{Micelle}}} + Ca(HCO_3)_2 \longrightarrow \underset{\text{(Normal Soil)}}{Ca^{++}\boxed{\text{Micelle}}} + 2H_2O + 2CO_2$$

iii) **Slaked lime**: It is produced by adding water to burnt lime.

$$CaO + H_2O \longrightarrow Ca(OH)_2$$

For every 56 Kg of calcium oxide, 18 kg of water is needed for proper slaking to yield 74 kg of hydrated or slaked lime. During this process considerable amount of heat is generated.

When slaked lime is added to an acid soil, it combines with carbonic acid to form calcium bicarbonate in solution.

$$\underset{\text{(Slaked lime)}}{Ca(OH)_2} + \underset{\text{(Carbonic acid)}}{2H_2CO_3} \longrightarrow \underset{\text{(Calcium biocarbonate)}}{Ca(HCO_3)_2} + \underset{\text{(Water)}}{2H_2O}$$

Calcium bicarbonate in solution reacts with the soil colloids and the calcium ions are adsorbed in the soil complex.

$$\underset{\text{(Soil colloids)}}{\begin{matrix} H^+ \\ H^+ \end{matrix} \boxed{\text{Micelle}}} + \underset{\text{(Calcium bicarbonate)}}{Ca(HCO_3)_2} = \underset{\text{(Soil colloid)}}{Ca^{++}\boxed{\text{Micelle}}} + \underset{\text{(Water)}}{2H_2O} + \underset{\text{(Carbon dioxide)}}{2CO_2}$$

As the reaction proceeds, CO_2 gas is freely evolved. The adsorption of calcium in the soil colloids raises the soil pH.

Efficiency of liming materials

The factors influence the effectiveness or efficiency of any liming materials are chemical composition, purity of the material and degree of fineness of the material.

i) **The chemical composition** of the material expressed in terms of calcium carbonate equivalent is known as neutralizing value (N.V.):

   Pure calcium carbonate is taken as the standard against which other liming materials are measured, and the neutralizing value of $CaCO_3$ is considered to be 100 per cent. $CaCO_3$ equivalent or N.V. of slaked lime or CaO can be calculated as follows:

$$CaCO_3 \text{ equivalent} = \frac{\text{Molecular wt. of } CaCO_3}{\text{Molecular wt. of } CaO} \times 100$$

$$= \frac{100}{56} \times 100$$

$$= 178.6$$

The same procedure can be followed to calculate neutralizing value of other liming materials. The $CaCO_3$ equivalent or neutralizing value of different liming materials are given in Table 4.

TABLE 4: Neutralizing value of commonly used liming materials

Material	Chemical Composition	Neutralizing value (%)
Limestone	$CaCO_3$	100
Quick lime	CaO	178
Slaked lime	$Ca(OH)_2$	135
Dolomite	$CaMg(CO_3)_2$	109
Basic slag	-	60-90

All the liming materials, if applied in proportion to their chemical equivalence and if of the same fineness have the same efficiency in long run. However, if quick results are required, the oxide and hydroxides are the best forms as they are very finely divided and dissolved immediately upon coming to contact with water.

ii) **Fineness of liming material**: The speed with which a liming material reacts with an acid soil depends upon its fineness. The finer the material is ground, the more rapidly it goes into solution. The liming materials, such as CaO, $Ca(OH)_2$, precipitate $CaCO_3$ and press mud are usually in powdery form and are thus very effective. The fine material has more surface area and hence, the change from carbonate to the soluble bicarbonate form or hydrolysis of silicate to hydroxides is quicker. The fine material can mix with greater volume of soil and as such neutralise acidity in a greater surface area. The finer particles are in more intimate contact with the soil particles than the coarser particles and hence, more effective.

The fineness is measured in terms of the ability of a material to pass through a sieve having 60 holes of equal size in one linear inch. Such a sieve is called a 60-mesh sieve and a material passing through such sieve is allotted 100 per cent efficiency rating. Efficiency ratings of various sized particles of liming material may be indicated as per such scale (Table 5) used in Ohio (USA).

TABLE 5: Efficiency rating of liming material

Nature of the material	Efficiency rating
1) Material passing through a 60- mesh sieve	100
2) Material passing through a 20 but not a 60-mesh sieve	60
3) Material passing through an 8 but not a 20-mesh sieve	20

The results of experiments conducted at Ranchi suggests that the minimum fineness of the limestone powder for liming of acid soils should be such that it passes through 10-20 mesh sieve. The data indicated that limestone powder passing through 10, 50 and 100 - mesh sieve were equally effective in increasing the yield of maize but the pH data showed that 100 - mesh particles (finest) raised the soil pH more quickly and to greater extent than the coarser particles. After the second year, the pH value of the soils receiving all the three mesh sized particles were more or less the same.

iii) **The purity of the material**: The effectiveness of liming material depends on its purity from the composition point of view. The purer the material, the more effective it is for neutralizing acidity. For example, if ground limestone contains 95 per cent of $CaCO_3$, the neutralizing value of the material will be 95 per cent.

Amount of lime to apply

The rate of liming would depend upon the following factors:

i) **The pH or intensity of the soil acidity**: For strong acid soils, more liming material will be required compared to slightly acidic soils.

ii) **The texture of the soil**: Sandy soils will require less lime compared to heavy soils like silt-loam and clay-loam.

iii) **Nature of crop**: Pulses and legumes benefit the most by the application of lime since they do not thrive on strongly acid soils and are also large consumers of lime.

iv) **The purity of liming material**: Purer the liming material, the less quantity is required.

v) **The chemical composition of liming material**: The CaO and $Ca(OH)_2$ have higher neutralizing values and hence, lime required for application will be less.

vi) **The finess of liming material**: The finer the liming material, the more effective will it be.

Although liming is essential to boost crop production in acid soils, the practice is not becoming popular with the farmers because of the initial heavy investment on liming. Some of the recent studies conducted in different parts of the country show that small amount of lime can improve the crop yield to a reasonable level in acid soils. It is observed that application of lime at 0.25 per cent of lime requirement shows appreciable direct and residual effect on crop production. This level of lime can alleviate the toxic effects of exchangeable Al and augment the uptake of P and Ca by the test crop (Prasad *et al.,* 1985).

It is reported that lime application as per lime requirement which varies from 2.88 to 5.00 t per ha have the highest yields of groundnut, soybean, chickpea and barley. However, **furrow application of lime** at 3 q per ha in soybean and 4 q ha in chickpea was at par with full dose of lime in respect of increase in crop yield. Results of the study suggest that lime may be used as fertilizer. In order to reduce the investment on liming, furrow application of smaller amount of lime may be advocated to achieve reasonable yields of lime responsive crops in acid soils (Rana *et al.,* 1985).

A large number of experiments conducted on crop response to liming and fertilization has revealed that application of 50 per cent NPK + Lime is more economical than recommended dose of 100 per cent NPK applied to the crops.

Frequency of liming

It depends on the following factors:

i) **Fineness of liming material**: Finer materials of lime are required to be applied at shorter intervals than coarser grained materials.
ii) **The point to which pH is to be raised**: The crop with optimum pH at 6 will require smaller dose at shorter intervals and so on.
iii) **Texture of the soil**: Light textured soils require liming in smaller dose and at lesser intervals than the clayey soils.
iv) **Climatic conditions**: Region with higher rainfall requires liming at shorter intervals than those receiving less rainfall.

Methods of lime application

Reaction of applied liming materials with soil colloids depends mainly on the nature of lime, soil texture and moisture conditions. When there is sufficient moisture, soil acidity is neutralized in a few days time. The following general principles should be observed in applying lime:

a) Lime should be applied before ploughing or applied on ploughed land and then disced or mixed into the soil.
b) In case of large quantity of lime are to be applied, it is advisable to apply small quantities two to three times rather than giving one heavy application.
c) On strong acid soils when larger dose of lime are required, half the quantity should be applied before ploughing and the remaining half applied and disced in after ploughing.
d) When lime requirement is 2 ton or less, the entire amount should be applied at one time, one month before sowing of the crops.
e) When only one application of lime is to be given, it should never be given before ploughing with a furrow turning plough.

ALTERNATIVE TO LIMING

An acceptable alternative to liming are:

i) To grow acid tolerant crop and cultivars laying emphasis on diversification.
ii) Agroforestry is an emerging option for resource management in acid soils of eastern hilly - agroecosystem utilizing the lowest 1/3 of the hill slopes put to agriculture crops, the middle 1/3 of hill slopes to horti-pastoral system and top 1/3 to agroforestry.
iii) Importance of integrated nutrient management system (INMS) applied to acid soils, such as organic manure, crop residues like rice straw, *rhizobia*, blue-green algae (BGA) and *azotobacter* cannot be under estimated.
iv) Organic matter complexes aluminum and manganese and thereby decreases their amount in soil solution. Where available large amount of organic matter, e.g. easily decomposable plant residues, could be incorporated into acid soils. The humified material should decrease the concentration of aluminum and manganese in the soil solution because of the formation of organo-complexes.

Species / Varieties Tolerant to Soil Acidity

The following crop species and genotypes identified as tolerant to soil acidity ($pH<5.5$) may be adopted to increase productivity of acid soils:

Acid Soil Region / State	Crop	Varieties
Assam	Rapseed	Varuna, Sonmukhi
	Summer greengram	K851, Sonmugu
Himachal Pradesh	Soybean	Bragg, Pb I, Harasoya
	Gobhi sarson	ONK I, Hoyala, HPN-3
Jharkhand	Blackgram	KU - 301
Kerala	Vegetable cowpea	Bhagyalakshmi
	Cowpea (Bush type)	V-16

Acid Soil Region / State	Crop	Varieties
Meghalaya	French bean	HUR - 15
Orissa	Groundnut	Smruti
	Pigeonpea	UPAS - 120
West Bengal	Mustard	Sanjukta, Pusa bold
	Wheat	J9107, PBW 343

Source: ICAR Network Project on Acid Soils. Reported by Sharma and Sarkar (2005)

The acid soil can also be suitably utilized by growing tolerant species for increasing the production of fuel, fodder and grazing resources (Singh, 1990).

Suitable Species

(i)	Grasses	*Pennisetum polystachyon, P. Pedicellatum, Paspalum notatum, Pennisetum clandestiannum* (Kikiya grass).
(ii)	Legumes	*Centrosema Pubesens, Desmodum, intortum, Stylosanthes guinesis.*
(iii)	Tress	*Ficus numeralis, Albizia chinensis, Morus serrata, Alnus nepalensis, Schima wallichi, Bucklandia pupulnea.*

EFFICIENT CROPS AND CROPPING SYSTEMS FOR ACID REGIONS

The existence of differential response of crops to liming in different parts of the country suggests that cropping pattern may be adjusted in limed soils to realise maximum benefits from the investment on liming.

Studies by Mandal *et al* (1969) at Ranchi revealed than cotton, groundnut, soybean and pigeonpea grew best only at pH 6.5 to 7.0 and as such may be grown in the first two years of lime application. This should be followed by medium responsive crops (pH 6.0 to 6.4) such as maize, sorghum, blackgram and greengram, etc. In the 5th or 6th year when the effect of lime is gradually reduced low responsive crops like rice, small millet and potatoes may be grown at pH below 6.0. Hence, a cropping following liming was suggested for rainfed uplands as follows:

1st year .... - Cotton
2nd year .... - Groundnut
3rd year .... - Pigeonpea
4th year ....- Maize
5th year ....- Upland rice
6th year .... - Rainy season potatoes

On the basis of trials on intercropping and successful growth of linseed in *rabi* (winter) season, the following modified crop sequences were suggested:

1st year .... - Groundnut - Linseed
2nd year .... - Maize + Pigeonpea
3rd year .... - Maize + Pigeonpea
4th year .... - Direct sown rice - Linseed
5th year .... - Rainy season potatoes - Linseed
6th year .... - Rainy season potatoes - Linseed

Based on experimental evidences efficient cropping systems (Table 6) for different acidic regions of the country have been identified which should be able to produce high yield on sustained basis (Srivastava, 1995).

TABLE 6: Efficient cropping systems in acid regions of India

(1) *Assam* (Barathakur et al., 1992)

	Flood Free	*Flood affected*
(A) Upper Assam	Autumn rice-Winter rice Winter rice-Pea Autumn rice-Mustard/Potato/blackgram/greengram Soybean-Mustard	Autumn rice-Pea/Mustard
(B) Central Assam	Autumn rice-Winter rice Winter rice-pea Summer rice Wheat- Autumn rice/soybean Autumn rice-Rabi pulses/mustard/potatoes	Autumn rice-Rai-Mustard/Pea/ Lintil/ Wheat
(C) Lower Assam	Autumn rice - winter rice Summer rice Wheat- Autumn rice/Soybean Autumn rice-Rabi pulse/mustard/potatoes	
(D) Southern Assam	Autumn rice - winter rice Summer rice	

(2) North Eastern Hill Regions (Planning Commission, GOI, 1989)

		Non-Irrigated		Irrigated	
		1st year	2nd year	1st year	2nd year
A.	Sub Zone I (Sikkim/ Darjeeling)	Rice/Maize/Fingermillet	Rice/Maize+Finger millet	Rice/Maize Mustard / wheat	Maize-Mustard
		Maize/Wheat/Mustard/Pulses	Maize/Millet/Finter-millet	Rice - Wheat	Rice-Wheat
		Rice/Maize-Mustard/Pulses	Potato-Potato	Potato-Potato	Potato- Potato
B.	Sub Zone II (Arunachal Pradesh, Nagaland, Manipur, Tripura, Mizoram, Meghalaya)	Maize +Pulses	Maize +Pulses	Jute-Rice	Jute-Rice
		Rice-Potato	Rice-Potato	Potato	Rice-Potato
		Rice-Potato - Tomato	Rice-Rice	Rice-Rice	Rice-Wheat/Pulses
		Rice- Potato-Vegetables	Potato Cole Vegetables	Mustard/ Wheat/ Potato	Oilseeds
		Jute-Blackgram/Mustard Sesamum/Groundnut	Sweet Potato-Rice Ahu rice/HYV rice	Rice-Rice-Rice	

(3) *Bihar Plains* (Roy Sharma *et al.*, 1992)

A. Zone II (Purnia district)	Jute (JRO 524)-Rice (Pusa 2-21); Jute (JRO 524)-Rice (Pusa 2-21)-wheat (Sonalika); Jute (JRO 524)-Blackgram (Navin)-Wheat (Sonalika); Jute (JRO 524)-Mustard (UP 70 SC -70)	Jute (JRO 524)-Wheat (HP 1102) Jute (JRO 524)-Potato (K. Sindhuri) Jute (JRO 524)-Peas (T 163)

(4) A. Bihar Plateau/ Jharkhand (Srivastava, 1989)

Sub-Zone	Rainfall	Climate	Soils	Existing Cropping patterns	Intensity of cropping, %	
					Existing	Proposed
III. North-Central Plateau (Chhotanagpur North & South Hill Plateau)	1296 (1100-1500)	Moist subhumid to dry subhumid	Red, sandy, red & yellow, patches of heavy grey soils of Rajmahal	a. Dhanbad, Giridih, Hazaribagh Ragi/Gondli-fallow Millet-Kulthi, Fallow-Niger, Rice + Arhar, Rice-fallow, Fallow-Til, Maize - fallow/Kulthi/ Niger Rice- lentil/gram / Linseed b. Deoghar, Godda, Sahebganj Fallow-Boro Rice, Maize-Gram / Lentil, Maize Potato- Cheena, Rice-Gram/Wheat/Barley/Fallow	110-115	150-300

Sub Zone Proposed Cropping Systems

		Unirrigated	Irrigated
(a)	**Districts:** Deoghar, Godda, Sahebganj and Dumka	Maize + Mung-Barley/Pulses/Oilseeds/ Gram/ Lentil, Rice-Wheat/Barley/Pulses / Oilseeds	Early Rice - Potato - vegetables Maize + Urd / Potato - vegetables Maize + Urd - Wheat Maize + Moong - Sugarcane Rice-Wheat-Rice Rice - Wheat Rice - Winter Maize -Moong
(b)	**Districts:** Dhanbad, Giridih, Hazaribagh & Plateau land of Santhal Parganas	Moong-Oilseeds/Pulses/Barley Moong - Toria Ragi - Safflower Jute for seed-fallow/Barley Arhar + Groundnut (1:2 or 2:6) Arhar + Rice (1:2 or 1:3) Maize - Linseed/Gram/Barley/Toria Tobacco-Fallow Cotton-Fallow Sweet Potato - Safflower Rice-Barley/Pulses/Oilseeds Rice + Paira crops	Maize + Urd-Potato-Wheat-Summer Moong Early maize-Early winter vegetables-Onion - ***Summer vegetables*** Maize-Wheat-Vegetables Rice-Wheat-Summer moong/Summer rice Maize-Potato-Wheat-Moong Jowar-Methi-Oats + Peas Vegetables all the year round Rice-Potato-Summer rice Rice-Winter maize - Summer rice Maize-Potato-rice Tobacco-Wheat/Pulses/Oilseeds-Moong Rice-Berseem

		Unirrigated	Irrigated
			Dinanath Grass-Cowpea-Fallow/Sugarcane/ oats-peas Cropping Intensity Existing: 110-115; Proposed: 150-300
	EASTERN PLATEAU: Existing (Chhotanagpur South, West Bengal Hills and Plateau)	Maize /Gondli-Fallow Maize - Fallow/Kulthi/Niger Fallow-Kulthi/Niger/Til	
	Five districts: Existing Palamau, Lohardaga, Gumla, Ranchi and Singhbhum	Rice + Arhar Rice-Fallow Rice-Gram/Khesari Rice-Wheat/Vegetables Maize-Vegetables	
	Five Districts: Proposed Palamau, Lohardaga, Gumla, Ranchi and Singhbhum	Maize - Pulses Cowpeas-Pulses/Oilseeds Cowpea-Dinanath grass -Fallow Moong/Soybean-linseed Groundnut-Linseed/Barley Jowar-Gram/Oats + Peas	Rice-Wheat-Rice Rice-Wheat-Vegetables/Moong Rice-Berseem + Oats/ Mustard Cotton-Wheat/Pulses Maize-Potato-Wheat-Moong Maize-Urd-Wheat-Moong Maize-Radish/Carrot/Turnip-Wheat/Barley/Pulses / Oilseeds Maize-Urd-Sugarcane/Potato/Mustard Groundnut- Mustard-Sugarcane + Moong (Feb. Planting) Rice-Potato-Rice

		Unirrigated	Irrigated
			Maize-Potato-Rice
			Maize + Urd- Potato-maize
			Dinanath grass + cowpea - sugarcane
B.	**Bihar Plateau/ Jharkhand** (Land situationwise) (Srivastava, 1989)		
	(i) Uplands (*Tanr* land)	Arhar + Maize (1:1)	
		Arhar + Rice (1:2 or 1:3)	
		Arhar + Groundnut (1:2 or 2:6)	
		Arhar + Urdbean (1: 2)	
		Maize + Soybean (1:2)	
		Ragi -Toria	
		Rainy season potato-Linseed / Toria	
		Jowar-Ratoon Jowar	
		Direct sown Rice / Mungbean/Urdbean-Toria /Linseed/Niger	
		Maize + Cowpea-Linseed/Safflower/Torial/Lentil Niger	
	(ii) Medium lands:	Rice-Linseed/Gram/Lentil/Barley	Rice-Wheat+Mustard Gram/Lentil (4:1)
		Groundnut-Linseed/Safflower	Rice-Potato
			Rice-Winter maize
			Jute for seed-Wheat
			Rice-Wheat-Moong

		Unirrigated	Irrigated
	(iii) *Lowlands: (Don lands)*	Rice-Linseed/Gram / Barley	Rice - Wheat Rice-Gram/Vegetables/ Lintil/Oilseeds Rice-zero tiled wheat-lentil Early rice-Wheat/Vegetables Jute for seed-potato-vegetables

(5) *Bengal* (Chatterjee *et. al.*, 1992)

	Major Agricultural Products	*Cropping systems*
Hill zone Darjeeling except Siliguri (pH 5-6) Tarai zone district of Jalpaiguri (pH 5-6) Old Alluvium zone, Birbhum parts of Burdwan and parts of Hoogly (pH 5.5 - 6.5)	Potatoes, Vegetables, Ginger, Cardmom, Orange, Rice, Maize and Tea Aus, Aman, Boro Rice, Jute, Wheat, Mustard, Puses, Potatoes, Oilseeds and Pineapple Rice, Jute, Wheat, Pulses, Oilseeds, Potatoes and Sugarcane	Autumn Rice/Maize-Potato/Wheat-Sesamum Autumn Rice/Jute-WinterRice- Sesamum Autumn Rice /Jute-Winter Rice-Wheat Autumn Rice /Jute-Winter Rice-Sesamum Autumn Rice/Jute-Winter Rice-Summer Rice Wheat/Potato/Rapeseed/Mustard-Sugarcane Winter Rice-Potato-Sesamum Autumn Rice/Jute-Potato-Sesamum Winter Rice-Potato- Sesamum Winter Rice-Potato/Wheat/Rapesed/ Mustard-Sesamum/Greengram

	Major Agricultural Products	*Cropping systems*
New Alluvium Zone Malda, West Dinajpur, Murshidabad (pH 5.5-6.5)	Rice-Jute, Wheat, Pulses, Oilseeds, Potatoes, Sugarcane, Barley, Sunflower	Autumn Rice/Jute/Greengram-Wheat / Rapseed/Mustard Autumn Rice/Jute-Winter Rice-Summer Rice/Sesamum Winter Rice- Wheat-Sesamum/Greengram Autumn Rice/Jute/Maize-Wheat-Greengram/ Sesamum Winter-Rice-Wheat
Lateritic Zone Purulia (pH 6.2-6.6), Bankura and parts of Midnapur (pH 5.5 - 6.5)	Rice, Maize, Millet and Pulses	Maize, Winter Rice-Wheat/Rapseed/Mustard Maize-Winter Rice-Sesamum/Greengram Autumn Rice-Winter Rice-Potato/Wheat or Summer rice Autumn Rice / Jute / Maize Winter Rice-Wheat/Potato-Greengram/ Sesamum

(6) *Orissa* (Lenka *et.al.*, 1992)

Zone		Cropping System
North-Western Plateau: (Sundargarh Dist. and Kuchinda of Sambalpur Dist.)	**Rainfed**	
	(i) Upland	
	Intercropping:	Pigeonpea+Groundnut/Greengram/Blacgram/Rice/Fingermillet
		Castor/Mazie/Sorghum +Greengram/Blackgram
		Rice+Mesta/Maize/Sorghum/Pigeonpea
	Sequence:	Minor millet-Horsegram/Greengram/Niger/Mustard
	cropping:	Maize/Jute/Sorghum/Rice-Horsegram / Greengram / Castor / Groundnut / Niger / Mustard/ Groundnut-Castor/ Horsegram / Niger / Greengram/ Blackgram /Horsegram /Mustard

Zone		Cropping System
	(ii) Medium land	Rice-Mustard/Linseed/Safflower
		Rice-Greengram/Lentil/Bengalgram/Field peas
		Rice-Bengalgram/Lentil/Field peas as under para sown crop.
	(iii) Lowland	Rice-Linseed/Lentil/Bengalgram/Field peas or these crops as undersown (*Chatta crop, Pyra*)
	Irrigated	
	(i) Upland:	Maize-Potatoes/Mustard-Sesamum/Vegetables
		Groundnut/Rice-Sugarcane-Groundnut/Sesamum/Greengram (2 yrs)
		Rice-Mustard/Wheat/Maize-Greengram/Vegetables
		Rice-Potato/Groundnut/Vegetables-sesamum/Greengram/ Vegetables
		Rice - Greengram-Groundnut
	(ii) Medium land:	Rice-Sugarcane-Sesamum
		Rice-Mustard/Potatoes/Wheat/Vegetable
		Jute-Rice-Greengram/Groundnut/Blackgram
	(iii) Lowland:	Rice-Mustard/Blackgram/Vegetables-Rice
		Jute-Rice-Blackgram-Sesamum
		Rice-Rice
North-Eastern Ghat (Phulbani, Aska, Parlakhemundi and Raygada)	**Rainfed**	
	(i) Upland	
	Intercropping:	Pigeonpea+Rice/Fingermillet/Maize/Groundnut/Greengram / Blackgram

Zone		Cropping System
		Maize/Sorghum/Cowpea/Greengram/Blackgram
		Castor + Greengram/Blackgram
	Sequence crop:	Maize - Horsegram
		Sorghum- Horsegram
		Greengram- Horsegram/Mustard/Niger
		Castor-Millet-Niger/ Horsegram
		Rice/Greengram/Blackgram-Horsegram/Niger/Mustard/ Groundnut
	(ii) Medium land:	Rice-Groundnut/Horsegram/Castor
		Rice-Greengram/Blackgram/Bengalgram/Safflower/Mustard
		Rice-Greengram/Blackgram (Chatta)
	(iii) Lowland:	Rice-Greengram/Blackgram/Field peas/Bengalgram/Mustard Linseed
West-Central Tableland (Sambalpur and Bargarh Sonepur and Belanger) and Mid-Central Table and (Dhenkanal, Angul, Athagarh and Sukunda block of Jaipur	**Rainfed**	
	(i) Upland	Rice-Field peas/Bengalgram (under sown Chatter)
	Intercropping:	Rice/Pigeonpea/Groundnut/Greengram/Blackgram/Fingermillet
		Maize + Cowpea/Greengram/Blackgram
	Sequence Crop:	Rice-Groundnut/Horsegram/Greengram/Blackgram/Mustard
		Groundnut-Horsegram/Caster
	(ii) Medium Land	Rice-Greengram/Blackgram or these crops under sown (*Pyra*) or Groundnut
	Irrigated	
	(i) Upland:	Maize/Rice-Groundnut/Wheat/Potato/Vegetables /

Zone		Cropping System
		Mustard-Sesamum
		Groundnut-Mustard/Maize-Vegetables
	(ii) Medium land:	Rice-Sugarcane-Groundnut/Sesamum/Greengram (2 yrs)
		Rice-Mustard/Potatoes/Wheat/Vegetables/Groundnut/ Greengram/Cowpea
		Rice-Mustard/Tomato/Potato/Rice
	(iii) Lowland	Rice-Rice
North-Eastern Coastal Plain: (Balasore, Bhadrak, Jaipur and Anandpur of Cuttack District) and East and South-Eastern Coastal plains (Cuttack, Jagatsinghpur, Puri, Khurda, Berhampur)	**Rainfed**	
	Inte(i) Upland *rcropping:*	Rice+Groundnut/Fingermillet/Pigeonpea/Blackgram/Greengram
		Maize + Greengram/Fingermillet/Mestal/Pigeonpea/Blackgram/ Rice for Ganjam district.
	Sequence Crop:	Groundnut-Horsegram
		Ragi/Rice-Groundnut/Horsegram/Greengram/ Mustard/Niger
		Maize-Horsegram/Greengram/Niger
		Jute/Mesta-Horsegram/Blackgram/Mustard
	(ii) Medium Land	Rice-Groundnut/Mustard/Blackgram/Coriander
		Jute-Blackgram/Coriander
		Rice-Blackgram/Greengram/ Coriander
	(iii) Lowland	Rice-Blackgram/Greengram/Linseed/Coriander
		Jute-Blackgram/Greengram or as *Chatta (Pyra)*

Zone		Cropping System
	Irrigated	
	(i) Upland:	Vegetables/Groundnut/Rice-Sugarcane-Sesamum (2 yrs)
		Rice-Potato/Groundnut/Wheat/Vegetables-Sesamum/
		Greengram/Vegetables
		Jute-Potato/Wheat/Groundnut/Rice-Blackgram/Ragi- Groundnut / Vegetables
	(ii) Medium land:	Rice-Groundnut/Wheat/Potato/Mustard/Vegetables/Sesamum-Rice
		Jute-Rice-Blackgram-Vegetables.
	(iii) Low land	Jute-Rice-Rice/Vegetables/Groundnut
		Rice-Rice-Rice
		Jute-Rice
		Rice-Field peas/Bengalgram (under sown *Chatta*)

(7) Madhya Pradesh /Chattisgarh (Bisen et al., 1992)

Zone	Existing		Recommended
A. Rice Zone			
(i) Chattisgarh plain	Rice-Rice	I.	***Bunded field (rainfed)***
	Rice-Utera (linseed)		Rice-Greengram/Blackgram/Linseed
	Rice-Wheat/Chickpea/Linseed		Rice-Chickpea/wheat
(ii) Bastar Plateau	Rice-Wheat/Chickpea	II.	***Irrigated***
	Rice-Linseed		Rice-Chickpea/Wheat
			Rice-Groundnut

Zone	Existing		Recommended
			Rice-Rice
			Rice-Maize
			Rice-Chickpea-Cowpea (Green)
			Rice-Potato-Maize
(iii) North hill region of Chhatisgarh	Rice-Fallow	III.	***Intercropping (rainfed)***
	Sorghum-Fallow		Pigeonpea + Sesamum
	Niger-Fallow		Pigeonpea + Kodo + Kulki
	Rice-Wheat/Chickpea/Linseed/		Wheat + Chickpea/Linseed
	Rape & Mustard		Sorghum + Pigeonpea/Blackgram
			Sesamum + Finger millet
			Finger millet + Kulki
B. Rice-Wheat Zone Kymore	Rice-Fallow/Utera	(i)	***Rainfed***
Plateau and Satpura hill	Rice-Wheat		Rice-Wheat/Chickpea
	Rice-Linseed		Rice-Linseed (utera)
			Maize-Rapeseed/Mustard
			Ramtil-Wheat
			Blackgram/Greengram-wheat
			Irrigated
			Rice-Wheat/Chickpea
			Rice-Lentil
			Rice-Chickpea + bhindi
			Rice-Potato-Maize (cobs) /Greengram

Zone	Existing		Recommended
			Rice-Wheat-Greengram/Maize
			Soybean-Wheat
			Greengram/Blackgram-Wheat/Chickpea
			Groundnut-Wheat/Rapeseed/Mustard
			Maize + Sesamum-Wheat /Chickpea/Linseed
			Intercropping
			Sorghum + Soybean
			Sorghum + Pigeonpea (3:2)
			Sorghum + Greengram/Blackgram
			Kodo + Pigeonpea
			Pigeonpea + Maize
			Maize + Blackgram/Greengram
C. Sorghum-Wheat Zone Part of Bundelkhand (Tikkamgarh, Chatterpur etc.)	Rice-Fallow	(i)	***Rainfed***
	Rice-Wheat/Chickpea/Linseed		Blackgram/Greengram-Wheat/Chickpea/Linseed
	Sorghum-Pulses		Sorghum-Chickpea/Linseed
			Groundnut-Wheat
			Maize-Chickpea
			Soybean-Chickpea
		(ii)	***Irrigated***
			Sorghum-Wheat/Chickpea
			Soybean-Wheat/Chickpea
			Groundnut-Wheat

Zone	Existing		Recommended
			Maize-Wheat-Cowpea/Greengram
			Sorghum-Wheat-Greengram
		(iii)	**Intercropping**
			Pigeonpea + Soybean (1:2)
			Pigeonpea + Greengram
			Soybean + Sorghum -Wheat + Chickpea
			Maize + Soybean

(8) Himachal Pradesh (Singh *et al.*, 1992)

Zone	Cropping Systems
Mid-hill subhumid (north-eastern part of Kangra, Sirmour, Chamba, Solan southern-eastern parts of Mandi, Kullu, Shimla)	Rice-Maize-Wheat/Barley/Chickpea/Pulses
High-Hill temperate wet (North-Western parts of Shimla, Kullu, Chamba, Mandi, Sirmour and Kangra)	Maize-Wheat/Potato

9. Kerala (Sreedharan and Nair, 1992)

Zone	Cropping Systems
Southern zone: (Trivendrum, Quilon, Alleppey and Kottayam)	Rice-Rice-Fallow/Vegetables Rice-Banana
Central zone: (Palghat, Trichur and Ernakulam excluding coastal saline tracts)	Rice-Rice-Fallow Rice-Rice-Rice (in command areas) Rice-Pulses-Fallow (Single crop wet land) Banana-Vegetables/Tropical Tubers
Northern zone: (Cannanore, Calicul and Malappuram)	Rice-Rice-Fallow Rice-Flood-Fallow (during second crop land in flooded) Flood fallow-Rice-Rice
High altitude zone: (Wynad and Idukki)	Rice-Fallow (Hill valleys and river banks) Rice-Banana Rice-Ginger Coffee-Ginger/Banana/Tropical (In uplands) Pepper (Undulating land) Cardamom (Steep slope and deep valley) Rubber (Recently cleared forest lands)
Onattukara zone (Quilon and Alleppey)	Rice-Rice-Sesamum (Wet lands) Fallow-Rice-Prawn culture (Throughout-Zone)
Kuttand and Kole zone	Kole zone is low lying along the coastal strip. Rice during summer (Pubja). Rest of the year zone gets flooded.

Zone	Cropping Systems
	Kuttand zone is 2-3 m below sea level and submerged under water for the major part of the year.
Pikkali zone	Rice (wet land) - Pawn culture
(Ernakulam)	Coconut (upland)
	Only one crop of rice is taken in the first season. During the rest of the period fields are submerged and pawn culture is practised.
Low rainfall zone	Rice-Rice-Fallow
(Attappady hills and eastern parts of Palghat)	

(10) Maharashtra (Umrani and Shinde, 1992)

Zone	Cropping Systems
Very high Rainfall zone with lateritic soil (Ratnagiri and part of Kolhapur with pH 5.5 to 6.5)	*Kharif*: Pearl millet, Groundnut, Sunflower, Pigeonpea
Very high rainfall zone with non-lateritic soil (Thana, Kalaba with pH 5.5 to 6.0 and southern tip fo Chanda with pH 5.5 to 6.8)	*Rabi*: Sorghum, Safflower, Chickpea, Wheat

(11) Tamilnadu (Palaniappan *et al.*, 1992)

Zone	Cropping Systems
Zone VI (High rainfall zone, annual 1469 mm) Part of Kanyakumari Dist.	***Major crops:*** Rice, Coconut, Topica ***Plantation cross:*** Arreanut and rubber
(Highly leached and acidic except the south-western coastal area where the soil is saline coastal alluvium)	***Crop sequence:*** Rice-Rice-Rice (Apr. - Aug.) (Sept. - Jan.) (Feb.- Apr.)
Zone VII (Hilly zone, annual rainfall 1000-5000 mm) Soil lateritic with pH around 5.0. Ootacomand and Coonoor Taluks of Nilgiri district and part of Madurai district)	***Major crops:*** *(i) Irrigated:* Rice-Cotton, Groundnut and Sugarcane (ii) *Rainfed:* Millets, Sesamum, cole, vegetables ***Plantation crops:*** Tea, Coffee, Spices, Banana ***Crop sequence:*** Potato-Wheat (May-Aug.) (Aug. - Jan.) Potato - Vegetables-Potato (Feb.-Jan.) (Jun.-Aug.) (Sept.-Dec.)
Karnataka **(Planning Commission, GOI, 1989)**	
Southern Plateau & Hill Region X	80% of Dryland farming with cropping intensity of 111%
Sub-Zone 1 (Bidar)	20% area of Sorghum, Pearl, millet, Finger millet should be diverted to soybean
Sub-Zone 5 (Mysore)	Rice/Pigeonpea-Cowpea (Forage)
Western Plains and Ghat Region XII	Rice/Finger millet - Pulses/Groundnut
Sub-Zone 4 (Shimoga)	Rice-Rice-Pulses; Rice-Maize/Ragi

(12) *Punjab* (Cheema *et al.*, 1992)

	Principle crops	Other important crops	Irrigation intensity	Cropping Intensity
i) Gurudaspur	Rice (31%)	Maize (3.7%) Sugarcane (1.3%)	87%	169%
ii) Part of Hoshiarpur	Maize (22.7%) Wheat (38.6%)	Rice (5.5%)	33%	145.6% Maize is grown mostly as rainfed. Rice and sugarcane is preferred in irrigated areas. In rainfed areas sesamum and mosh during *Kharif* and pea as pulse during *rabi* are taken.

Some of the points that have to be remembered while dealing with acid soils:

1) Rice, millets, oats and linseed tolerate high acidity.
2) High lime responsive crops are pigeonpea, soybean, and cotton.
3) Acid soils without amendments are not suitable for growing wheat.
4) N should be applied in the NH_4^+ or amide form. Nitrate form should be avoided specially under lowland condition.
5) Liming helps in raising the base saturation of the soil and inactivates Al, Fe and Mn in soil solution. It helps to minimise P fixation and stimulates microbial activity.

REFERENCES

Barathakur, B. C., Thakur, A.C., Choudhary, A. K. Dutta, T. C. and Kalita, D. R. 1992. Efficient cropping system in Assam. Proc. National Symp. on Efficient Cropping System Zones of India held at Bangalore, Jan. 8-10, 1988. *Indian Society of Agronomy*, N. Delhi, pp: 39-74.

Barathakur, H. P. 1990. Management of Acid soils. In Agronomic Research Towards Sustainable Agriculture (**eds**. K. N. Singh and R. P. Singh) Indian Society of Agronomy, N. Delhi, pp:119-130.

Bhan, H. L., Mc Neal, B. N. and Oconner, A. G. 1979. Soil Chemistry, John Willey and Sons, New York (USA), pp:195.

Bhaumik, H. D. and Donahue, R. H. 1966. Soil acidity and the use of lime in India. Ministry of Food and Agriculture, Govt of India.

Bisen, C. R., Dubey, S. K. and Gogulwar, N. M. 1992. Efficient cropping systems in Madhya Pradesh. Proc., National symp. on Efficient Cropping System Zones of India held at Bangalore Jan. 8-10, 1988. *Indian Society of Agronomy*, N. Delhi, p. 299-342.

Chakraborty, M., Chakravarty, B. and Mukherjee, S. K. 1961. Liming for crop production in India. *Indian Society of Soil Science* Bulletin, 2.

Chatterjee, B. N., Mukhopadhyay, S. K. and Khan, S. A. 1992. Efficient cropping systems in sub-humid zones of India - West Bengal. Proc. National Symp. on Efficient Cropping Systems Zones of India held at Bangalore, Jan. 8-10, 1988, *India Society of Agronomy*, N. Delhi, p. 519-542.

Cheema, S. S., Kalan, J. S. and Grewal, S. S. 1992. Efficient cropping systems in Punjab, Proc. National Symp. held at Bangalore, Jan. 8-10, 1988 on Efficient Cropping System Zones of India, p. 411-438.

Gupta, R. K., Rai, R. N., Singh, R. D. and Prasad, R. N. 1989. Journal Indian Society of Soil Science, 37: 126.

Jackson, W.A. 1967. Physiological effects of soil acidity. *In* R. W. Pearson and F. Adims (*ed*). Soil Acidity and Liming, American Society of Agronomy, Madison, Wisconsin.

Kamprath, E.J. 1970. Proceedings of Soil Science Society of America, 34: 252-254.

Kamprath, E. 1990. Soil acidity in well drained soils of tropics as a constraint to crop production. In soil related constraints to food production in tropics. IRRI, Manila, Philippines.

Kanwar, J. S. and Joshi, M.D. 1964. Research Bulletin No. 1 (Ludhiana).

Lenka, D., Nayak, B. C. and Bhol, B. 1992. Efficient cropping systems in different agro-climatic zones of Orissa. In Efficient Cropping System Zones of India. Proc. of National Symp. held at Bangalore, Jan. 8-10, 1988. Indian Society of Agronomy N. Delhi, pp: 381-410.

Mandal, S. C., Das, S. K. and Mukherjee, H. N. 1955. *Journal of Indian Society of Soil Science*. 3:71.

Mandal, S. C. 1962. Journal of Ranchi Agriculture College and Research Institute, 2:39.

Mandal, S.C. and Jha, K. K. 1965. Proceedings of Seminar of Fertilizer and Crop Production in Eastern India, pp:89.

Mandal, S. C., Sinha, H., Prasad, C. R. and Ali, M.A. 1966. Journal of Indian Society of Soil Science, 14: 127.

Mandal, S. C. and Jha, K. K. 1969. Improvement of acid soils in Bihar by Liming. Technical Bulletin No. 1 Agriculture Research Institute. Kanke, Ranchi.

Mandal, S. C., Jha, P.P., Rao, M. S. 1969. Crop adjustments in the Plateau region of Bihar. Technical Bulletin No. 3, Agriculture Research Institute, Kanke, Ranchi.

Mandal, S. C. and Sinha, H. 1969. Soils of Bihar and their fertility (ii) Acid soils and their management. Fertilizer News, 14: 1-4.

Mandal, S.C. and Jha, K. K. 1970. Technical Bulletin No 5. Agricultural Research Institute, Kanke, Ranchi.

Mandal, S. C. 1971. Presidential Address of 58th Indian Science Congress, Banglore.

Mandal, S.C., Sinha, M. K. and Sinha, H. 1975. Acid soils of India and liming. *Technical Bulletin*, 51, ICAR, New Delhi.

Mandal, S.C. 1995. Nutrient Dynamics and Classification of Acid Soils. **In**. Acid Soil Management (*eds.* M.A. Mohsin, A. K. Sarkar and B. S. Mathur). Kalyani Publishers, Ludhiana -New Delhi, pp: 83-91.

Marion, G. M., Mendricks, D. M., Dutt, G. R. and Fuller, W. H. 1976. Soil Science, 121:76.

Mathur, B. S. and Srivastava, L. L. 1979. Increasing productivity of acid soils in eastern region-Bihar. *Proc. Symp. Increasing Productivity of Acid Soils in NE Region*, FAI, Patna, p. 90-100.

Mehlieh, A. 1933. Proceedings of Soil Science Society of America, 3:162.

Panda, N. 1979. Proc. Sym. Increasing Productivity of Acid Soils of NE region, FAI, Patna, p. 117.

Panda, N. 1987. *Journal of Indian Society of Soil Science*, 35: 568. c.f. Management of Acid Soils by Barathakur, H. P. in Agronomic Research towards sustainable Agriculture. Indian Society of Agronomy, IARI, New Delhi, p. 119-130.

Panda, N. 2000. Management of acid soils for sustainable agricultural production. International Conference on Managing Natural Resources for Sustained agricultural production in 21st Century held on Feb. 14-18, 2000, New Delhi, India. Extended Summaries - Invited Papers, 1: 43.

Patel, C. S. and Awasthi, R. P. (1992). Cropping Systems - Problems and Approach in North -Eastern Hill regions. In Efficient Cropping Systems Zones of India held at Bangalore. Jan. 8-10, 1988. *Indian Society of Agronomy*, New Delhi, p. 363-380.

Planning Commission (1989). Agro-Climatic Regional Planning- An Overview Planning Commission, Govt. of India, New Delhi.

Ponnamperuma, F.N., Bradified, R. and Peech, M. (1955). Physiological diseases of rice attributable to iron toxicity. *Nature*, 175: 265.

Roy, H. K. (1967). M.Sc. Ag. Thesis, Ranchi University, Bihar.

Prasad, R. N., Patiram and Munna Ram (1985). Journal of Indian Society of Soil Science, 33:469.

Pratt and Blair, F. L. 1962. California Agriculture, 15:13

Rich, C. I., 1955. Bulletin. Virginia Agricultural Experimentation Station, 475:2.

Reira, A. 1946. Journal of Agricultural University. R. R., 30: 184-185.

Rana, N. K. Singh, H. and Lal, S. 1985. Journal of Indian Society of Soil Science, 33:328.

Raychaudhury, S. P., Agrawal, R. R., Dutta Biswas, N. R., Gupta, S. P. and Thomas, P. K. 1963. Soils of India, ICAR, New Delhi.

Raychoudhary, S. P. 1974. Acid Soils in India. Proceedings. Autumn Institute, ICAR, Ranchi, Nov 4-Dec. 3, 1974.

Roy Sharma, R. P., Mishra, S.S., Sharma, H. M. Prasad, U. K. Singh, A.B. and Sinha, K.K. 1992. Efficient Cropping System Zones of Bihar. In Efficient

Cropping System Zones of India held at Bangalore, Jan. 8-10, 1988. Indian Society of Agronomy, New Delhi, p. 75-118.

Sahu, B. N. 1965. *Journal of Indian Society of Soil Science*, 13: 24. *In* Management of Acid Soils by H. P. Barathakur. Agronomic Research towards Sustainable Agriculture, Indian Society of Agronomy, IARI, New Delhi.

Sharma, P.D. and Sarkar, A.K. 2005. Managing Acid Soils for Enhancing Productivity. National Resource Management Division, ICAR, Krishi Anusandhan Bhawan-II, Pusa Campus, New Delhi. Technical Bulletin, pp:19.

Singh, C. M., Sharma, H. L. and Moorty, T.V. 1992. Efficient Cropping Systems in Himachal Pradesh. In Efficient Cropping System Zones of Inda. Proc. National Symp. held at Bangalore, Jan. 8-10, 1988. Indian Society of Agronomy, p. 191-126.

Singh, P. 1990. Forage Production for Sustainable Mixed Farming. *In*. Agronomic Research Towards Sustainable Agriculture (*eds.* K. N. Singh and R. P. Singh). Indian Society of Agronomy, ICAR, New Delhi, India, pp:108.

Sreedharan, C. and Nair, M.A. 1992. Efficient cropping systems in Agro-climatic Zones of Kerala. In Efficient cropping System Zones of India. Proc. National Symp. held at Bangalore, Jan. 8-10, 1988. Indian Society of Agronomy, pp: 281-298.

Srivastava, V.C. 1989. Crops and Cropping Systems in Bihar Plateau - Planning and Strategy, Department of Agronomy, Birsa Agricultural University, Ranchi.

Srivastava, V. C. 1995. Crops and Cropping Systems. *In* Acid Soil Management (eds. Mohsin, M. A. Sarkar, A. K. and Mathur, B. S.) Kalyani Publishers, Ludhiana, India, pp:359-382.

Tanaka, A. and Navasero, S. A. 1966. Aluminium toxicity of the rice plant under water culture conditions. Soil Science & Plant Nutrition 12: 55-60.

Tanaka, A. and Yoshida, S. 1970. Nutritional disorders of the rice plant in Asia. IRRI, Manila, Philippines p. 51.

Umrani, N. K. and Shinde, S. M. 1992. Efficient Cropping Systems in Scarcity Zones of Maharashtra. In Efficient Cropping System Zones of India. *Proc. National Symp.* held at Bangalore, Jan. 8-10, 1988, p. 343-362.

Varade, S. B. and Ghildayal, B. P. 1967. Journal of Indian Society of Soil Science, 15:157.

Woodruff, C. M. 1948. Soil Science, 66: 33

Yadva, J.S.P and Banerjee, S.P. 1968. Journal of Indian Society of Soil Science, 16:32.

Yawalkar, K. S. and Agarwal, J. P. 1962. Manures and Fertilizers. Agri-Horticultural Publishing House, Nagpur, India.

CHAPTER 3

Salt Affected Soils (Saline and Alkali)

Intensification of agriculture supported by irrigation, an inevitable consequence of the compulsion to produce more, has put an enormous burden on the natural resources, particularly soils. This has rendered many productive soils unfertile. Under normal condition, soils affect yield through three primary factors of moisture availability, nutrient availability, and physical conditions. A fourth factor of presence of excess salt due to accumulation of chemical components in the soil are inhibitive to plant growth. Increasing depletion of essential nutrients from soil, waterlogging, high water table and salinisation are threatening sustainability of crop yields in irrigated areas with poor drainage in several parts of the world. Hence, strategies for reclamation and management of such soils are of vital importance.

DISTRIBUTION OF SALT AFFECTED SOILS

Approximately 952.2 million ha of soils on earth's land surface are salt affected. This constitutes to nearly 7 per cent of the total land area and nearly 33 per cent of potential arable land (Chhabra, 2002). Out of 240 million ha that are not strongly saline (Massoud, 1974), about 49 million ha are in the humid regions of South and Southest Asia ; 27 million ha of these are coastal saline soils. The actual area of saline soils in the regions may be considerably higher than estimated - vast areas of coastal saline soils not reported are under mangrove vegetation.

In India, saline and alkali soils occur most commonly, but not exclusively under arid and semi-arid climatic regions. There are certain fields which contain excessive concentration of soluble salts, or exchangeable sodium, or both. There are four major tracts where saline and alkali soils are commonly met. They are (i) the arid alluvial tract of Punjab and Haryana, (ii) the semi-arid alluvial region of Uttar Pradesh, (iii) the arid and semi-arid tracts of Deccan, consisting principally of the irrigated soils, and (iv) coastal alluvium of littoral origin. A combination of factors which are mainly geological, climatic and hydrological in nature is usually involved in the formation of saline and alkali soils. According to Singh (1992), out of 8.373 million ha of salt affected soils, 2.359 million ha are alkali, 3.829 million ha are saline while the rest 2.185 million ha are coastal saline soils. These soils are mostly found in five states of Haryana, Punjab, Rajasthan, Gujarat and Andhra Pradesh accounts for 48 per cent of total salt affected soils of the country. While the penisular region contributes 1.2 million ha alkali and 1.0 million ha waterlogged saline soils, the coastal region of the country is affected by 2.1 million ha of saline to acid sulphate soils. Yadav (1996) estimated that area affected in India 6 million ha by salinisation and 4 million ha by sodication. Chahabra and Kamra (2000) reported that salt affected soils cover an area of about 8.57 million ha, almost half are in canal commands. Broad Groups of salt affected soils in India are shown in table 6.

TABLE 6: Geographical distribution of salt affected soils

Sl. No.	Group	States	Approximate Area (million ha)
1.	Coastal salt-affected soils		
	(a) arid region	Gujarat	0.714
	(b) humid region	West Bengal	1.394
	(c) acid salt affected soils	Orissa, Andhra Pradesh, Tamil Nadu, Kerala	0.016

Sl. No.	Group	States	Approximate Area (million ha)
2.	Salt affected soils of the medium and deep black soil regions	Karnatka, Madhya Pradesh, Maharastra, Andhra Pradesh	1.420
3.	Salt affected sols of the arid and semi-arid regions	Gujarat, Rajasthan, Punjab, Haryana, Uttar Pradesh	1.000
4.	Alkali soils of the Indo - Gangetic plains	Haryana, Punjab, Uttar Pradesh, Bihar, Rajasthan Madhya Pradesh	2.500

Source: Abrol and Bhumbla (1971)

Further large areas are getting saline every year on account of adoption of certain irrigation and agricultural practices and absence of effective drainage.

TERMS AND DEFINITIONS

Alkali Soils - A soil that contains sufficient exchangeable sodium to interfere with the growth of most crop plants, either with or without appreciable quantities of soluble salts. (See Saline-Alkali and Nonsaline-Alkali Soil)

Non-saline-Alkali Soil - A soil which contains sufficient exchangeable sodium to interfere with the growth of most crop plants and does not contain appreciable quantities of soluble salts. The exchangeable sodium percentage is greater than 15, the conductivity of the saturation extract is less than 4 ds per m and the pH of saturated soil paste usually ranges between 8.2 and 10.

Saline-Alkali Soil - A soil containing sufficient exchangeable sodium to interfere with the growth of most crop plants and containing appreciable quantities of soluble salts. The exchangeable-sodium-percentage is greater than 15 and the conductivity of the saturation extract is greater than 4 ds per m. The pH of the saturated soils paste is usually less than 8.2

Saline Soil - A nonalkali soil containing soluble salts in such quantities that they interference with the growth of most crop plants.

The conductivity of the saturation extract is greater than 4 ds per m, the exchangeable -sodium- percentage is less than 15, and the pH of the saturated soils paste is usually less than 8.2.

Alkalization - A process of accumulation of salts in the soil.

Exchangeable-sodium-percentage - This term indicates the degree of saturation of the soil exchange complex with sodium and is defined as follows:

$$\text{ESP} = \frac{\text{Exchangeable sodium (m.e. per 100 g. soil)}}{\text{CAtion exchange capacity (m.e. per 100 g. soil)}} \times 100$$

Soluble-sodium-percentage - The proportion of sodium ions in solution in relation to the total cation concentration, defined as follows:

$$\text{SSP} = \frac{\text{Soluble sodium concentration (m.e. per liter)}}{\text{Total salt concentration (m.e. per liter)}} \times 100$$

CLASSIFICATION

Variations in both quantity and nature of salts present in the surface horizon of soils usually form the basis for grouping soils into different categories (Singh and Chhabra, 1990). The most prevailing system of grouping all such soils has been the one followed by the U.S. Salinity Laboratory workers (Richards, 1954). This system, however, ignores the nature of salts present in the soil. Based on pH, exchangeable sodium, quantum and nature of salts, these soils have been classified into two broad groups of alkali and saline soils as given in Table 7. The term saline-alkali applies to soils which are both saline and alkali.

TABLE 7: Classification of salt affected soils

Type	Conductivity ds/m	Soil pH	Exchangeable sodium percentage	Other features Chief ions
A. Saline Soil	> 4.0	< 8.2	< 15	Chloride and sulphate with little bicarbonates. Physical conditions satisfactory.
B. Alkali soil	< 4.0	> 8.5	> 15	Bicarbonate with sodium. Very poor physical condition.
C. Saline-Alkali Soil	> 4.0	> 8.2	> 15	Carbonates with some chlorides and sulphates.

Soils containing excess concentration of salts have a pH more than 7.0. Such soils are sub-divide into the following four classes on the basis of pH range:

Soil type	**pH**
Mildly alkaline	8.4 - 7.8
Moderately alkaline	7.9 - 8.4
Strongly alkaline	8.5 - 9.0
Very Strongly alkaline	9.1 and higher

DIAGNOSIS OF THE PROBLEM

Diagnosis of the nature of soil problem is very essential for making appropriate recommendations for the reclamation and/or management of salt affected soils. Diagnosis for rapid screening and for a more complete understanding have been discussed by Abrol and Fireman (1977) as given below. Abrol *et al.* (1980) have suggested that pH of 8.2 instead of 8.5 may be a more realistic diagnostic criteria for distinguishing sodic from non-sodic soils.

i) **For Rapid Screening**, only pH and electrical conductivity of soil are usually determined and these are made on 1:2 soil-water mixtures. These tests give only an indication of the nature of problem.

pH greater than 9.0 usually indicates that the soil is likely to have an alkali problem. Higher the pH, the more likely an alkali problem and higher, usually, is the alkali status. In many alkali soils occurring in the Indo- Gangetic alluvial plains in the states of Haryana, Punjab and Uttar Pradesh, a fair good correlation has been found to exist between pH (1:2) and exchangeable sodium percentage and therefore, pH and the gypsum requirement of the soil. The relationship is only approximate and not valid for all other states.

EC greater than 1.0 ds per m in 1:2 soil : water mixture indicates that soil has salinity problem. If the salinity is the problem, the pH of the soil in 1:2 mixture is invariably less than 9.0 and more often less than 8.5. Leaching alone is required for reclamation.

ii) **For a more complete understanding** of the nature of the problem, it is preferable to evaluate pH on saturated soil paste and EC on a saturation extract. Determination of exchangeable sodium percentage, hydraulic conductivity and the gypsum requirement assist in making sound economic recommendations for reclamation of alkali/salinity problems.

a) pH of the soil paste greater than 8.2 indicates an alkali problem. A value less than 8.2 usually, but not always, indicates the absence of serious alkali problem.

b) EC of saturation extract greater than about 4 to 6 ds per m indicates a salinity problem. If the salinity is the chief problem, pH of the saturation paste is nearly always less than 8.2.

c) pH of saturation paste greater than 8.2 and also the EC greater than 4 to 6 ds per m, indicate the presence of both salts and alkali. In this case, the chief problem probably is that of high alkali status, and more complete alkali diagnosis and reclamation procedure are recommended.

d) Low hydraulic conductivity (0.1 mm per hour or below) usually confirms that the soil has alkali problem. High hydraulic conductivity is invariably associated with absence of any serious alkali problem; the possibility of a salinity problem still exists.

e) Exchangeable sodium percentage values higher than about 15 impart to most soils poor physical properties and related nutritional problems. The quantity of an amendment required to reclaim an alkali soil depends on the amount of exchangeable sodium that needs replacement by calcium. Replacement of 1 m.e. of adsorbed sodium from the surface 15 cm of soil requires approximately 1.9 tonnes pure gypsum per ha.

A. SALINE SOILS

Saline soils were originally called 'white alkali' soils. Russians term them as "solanchak' soils. Now the terms is gradually changing to saline soils. It is a non-alkali soil that contains soluble salts in such quantities that interfere with the growth of most crop plants.

Symptom Shown by Saline Soils

Such soils are often recognised by:

i) the presence of white salt crust or efflorescence,
ii) damp oily looking surface devoid of vegetation,
iii) stunted growth of plant,
iv) deep blue-green foliage with considerable variability in size, and/or
v) tip burn and firing of the margin of leaves

Measures of salinity

Salinity can be measured by establishing the soluble salts in per cent weight of the total soil or by electrical conductivity. But the best method which can be correlated to the plant growth is by electrical conductivity method. The total soluble salts by weight does not give any definite idea. The osmotic pressure which is responsible for poor plant growth, differs in soils of different texture and of different water holding capacity or saturation percentage. At the field capacity (moisture content at 1/3 bar), a clayey soil will have more moisture than a sandy soil, and with the same amount of soluble salt a sandy soil will have a higher salt concentration than a clayey soil at field capacity. For example, a soil with saturation percentage 75 will grow satisfactorily most crops at a salt content of 0.2 per cent, whereas a sandy soil with the same salt content will grow only very salt tolerant plants. Thus, crops that grow with water standing in the field, like rice are affected very little by high concentration of salts.

Characteristics

Chemical characteristics

i) Saline soils are characterised by pH <8.2, ESP < 15 and EC>4 ds/m.
ii) Presence of excessive amounts of soluble salts, mainly cl and SO_4 of Na, Ca and Mg limit optimal crop growth due to toxic effect of salts and increased osomotic stress which physiologically inhibits the availability of soil water. Na seldom comprises more than half (50%) of the soluble cations. The relative amounts of Ca and Mg in the soil solution may vary considerably. Soluble and exchangeable K are only minor constituents but may occasional occur in excess. Boron is in very small amounts generally less than 0.7 ppm but sometimes in toxic concentration (more than 1.5 ppm).
iii) Though they predominant chief anions are chloride and sulphate but sometimes nitrates of Na, Ca and Mg may be present. Small quantity of bicarbonate may occur.
iv) Presence of chloride, nitrate and sulphate of potassium in low amount but sometimes may be the predominant salts.

v) In those soils where the amount of bicarbonates is heavy, there may be a deficiency of Fe, Cu and Zn, as they become less soluble.

vi) In presence of high amounts of K salts, the amount of Ca ions is lowered resulting the depressed uptake of Ca.

Physical characteristics

i) While efflorescence on the surface known as white alkali.

ii) The colloids in some saline soils are highly flocculated because of the presence of excess salts and the absence of appreciable amount of exchangeable Na, thus providing a favourable structure and generally improved permeability to water and air and consequently, the excessive salts can be readily be leached below the root zone with irrigation water.

iii) Such soils have good aggregation.

Microbiological activity

High concentration of salts causes high osomotic pressure resulting less microbial activity.

Development of saline soils

Saline soils occur for the most part in regions of arid and semi-arid climates. They, however, also occur in semi-humid to humid regions in basin shaped lands and flat lands with restricted drainage or lands which are inundated by sea water.

The favourable conditions for formation of saline soils are: (i) a high water table with a fairly high salt concentration (ii) a high temperature, and (iii) a low rainfall. Consequently, the soils of dry regions tend to become increasingly saline as long as the groundwater remains within capillary reach of the evaporation from the surface of the soil.

The immediate source of salts in saline soils can be parent material, irrigation water, shallow brackish groundwater, surface or underground saline water or wind.

The soluble salts that accumulate in soils consist principally of

i) cations of calcium, magnesium and sodium, and
ii) anions of chloride and sulphate in large quantities, and bicarbonate, carbonate and nitrate ions in smaller quantities.

The proportion of various salts present in the soils make them saline or alkaline.

The formation of saline soils under different conditions are briefly described below:

Arid and Semi-arid regions

i) In these regions, the percolating water is unable to reach the ground water table. Thus, soluble salts liberated by the weathering of rocks and minerals, are not washed away, brought to the surface by capillary rise where they are deposited. This goes on year after year and concentration of salts increases.
ii) During rainy season, soluble salts near the surface may move downward to the lower soil layers but after the rainy seasons, the intense evaporation brings the salts back to surface.
iii) The ground water of arid regions usually contain considerable quantities of soluble salts. If the water table is high, large amounts of water move to the surface by capillary action and are evaporated, leaving an ever-increasing accumulation of soluble salts.
iv) The regions near the sea, lot of salt is blown by wind year after year and get deposited on the lands. Due to low rainfall they are not washed back to sea and thus, add to the salinity of the land.

Basin shaped lands

These lands get washing from surrounding lands also accumulate soluble salts as the evaporation is more than rainfall. Salinity develops more rapidly when these lands have also restricted drainage.

Humid and Semi-humid regions

In these lands water is lost mostly due to evaporation, the restricted drainage, as mentioned above for basin shaped land, may cause salinity. In river deltas or low laying areas near the sea which get sea water during tides, salt water may accumulate and enrich the soil with salt. The saline soils of Sunderbans of the Ganga delta and other reverine deltas have developed in this way.

Salt Accumulation in Irrigated Soils

Under irrigation, the saline soils have developed by one or more of several means, as follows:

i) When irrigation water has a high salt content and this add to the salinity of soil due to high water table, low permeability due to bad structure and heavier texture of soils, and pan or indurated layers, like clay pan or silica layer and hence, the irrigation water finds no escape but causing waterlogging and water may be lost due to evaporation enriching the soil with soluble salts.
ii) Poor drainage keeps the salt in the surface of soil and salt content in applied irrigation water is deposited in the root zone due to evaporation or consumptive use and prevents the leaching of salts.
iii) Excessive application of water have raised ground water level sufficiently to permit concentration of salts from saline ground water through evaporation.
iv) Seepage from leaky canals and lateral channels which run at a higher level has resulted in high water table.

v) When the use of irrigation water is erratic, i.e. flooding at one time followed by intense drought. This would leave salts in the root-zone due to limited water supply.
vi) High cropping intensity coupled with injudicious on-farm water use and replacement of low water requiring crops by crops of high water demand gradually result in high water table associated with salinization of soils.

Salinity and Plant growth

Plant growth on saline soils depends on several plant and environmental factors. Salt injury may be due to osomotic stress, specific ion effects, ionic antagonisms, toxicities caused by ions released by cation exchange, and effects of ionic activities. The effect of soil salinity may manifest itself mainly in the following ways:

i) The physical effects of salts increase the osomotic pressure of the soil solution, the rooting medium and thereby decreasing water entry and availability to the plants. As osomotic pressure increases, the rate and per cent of germination and plant growth decrease.
ii) Accumulation of certain ions, like Na^+, Mg^{++} K^+, SO_4^{--}, HCO_3^-, NO_3^- and possibly boron in toxic concentration in plant tissues, hampers plant growth and development. For example, high concentration of sulphate generally decreases the uptake of calcium while promoting the uptake of sodium inducing sodium toxicity to susceptible species. Sufficiently high concentration of certain ions may be toxic to embryo or seedlings inducing abnormalities in the growth and development of plants. Salinity also tends to increase the incidence of certain diseases, like blossom-end rot tomatoes.
iii) Soil salinity inhibits the growth of plant by altering the plant mineral nutrition.

According to Maas and Hoffman (1977), crop yield decreases markedly with increase in salt concentration, but the threshold concentration and the rate of yield decrease vary with the species. They describe the effect of salt concentration on yield by equation:

$$y = 100 - B\,(EC_e - A)$$

Where, y is relative yield, A is salinity threshold value, B is the yield decrease per unit of salinity increase.

Strongly saline soils are barren. Less strongly saline soils in arid areas are characterized by patchy growth of halophytic grasses and shrubs (FAO- UNESCO, 1973), whereas coastal saline soils of humid tropics and sub-tropics are characterised by the presence of mangrove species (Chapman, 1976). Most plants suffer salt injury at EC values exceeding 4 mmhos per cm at 25°C (US salinity Laboratory staff, 1954).

Rice is moderately susceptible to salt injury (Mass and Hoffman, 1977). The symptom of salt injury in rice are stunted growth, rolling of leaves, white leaf tips, white blotches in the leaves and drying of the older leaves.

Management of Saline Soils

Provision of adequate surface and sub-surface drainage to lower water table below the critical depth and to facilitate leaching of excess soluble salts from the root zone are fundamental to the reclamation of saline soils. Management of saline soils include reclamation methods and adoption of agronomic practices conforming to the need for keeping root environment at minimum salt levels. It is easy to achieve some of these objectives in areas where fresh water is available and there is a suitable drainage outlet for disposal of excess water.

1. Reclamation of Saline Soils

The reclamation of saline soils depends more or less on the factors which have contributed towards the development of salinity.

1) **Leaching excess salts**: Leaching involves removing excess salts to reduce root-zone salinity to a desired level. Saline soils in which the soluble salt contain appreciable amount of calcium and magnesium do not develop into alkali soils by the action of leaching water. Leaching can be accomplished by ponding water on the soil surface and allowing the water to stand for a week or fortnight and then it is allowed to escape. Excess soluble salts, thus, dissolved in water are either leached to lower layers beyond the root-zone of shallow rooted crop or lost in drainage. Practical methods to achieve maximum leaching efficiency include intermittent ponding or sprinkle irrigation. It helps in saving 30 to 70 per cent of water required for leaching through continuous ponding. However, it takes nearly 40 per cent more time for leaching as compared with continuous ponding. The treatments are repeated 2 to 3 times to reclaim highly saline soils. The best time for leaching salts would be when the ground water-table is deep or receding. In the Indo-gangatic plains, the water table is deepest in June before onset of monsoon.

   In highly permeable soils, it is not always possible to leach the salt beyond the root-zone because they do not permit the entry of water. Addition of gypsum considerably facilitate water penetration and therefore, salt leaching.

   Saline soils, which do not contain sufficient soluble calcium salt or reserve calcium carbonate, would be converted to alkaline soils upon leaching with water. This would increase difficulties instead of reclaiming. In such cases, the addition of calcium salts, preferably gypsum or limestone is beneficial. In the presence of calcium sulphate, the quantity of calcium is increased in soil solution, with the result that calcium displaces sodium from the exchange complex and a favourable physical change in the soil takes place. The amount of gypsum to be applied would depend upon the characteristics of the soil. According to Dakshinamurty *et.al.* (1973) in highly deteriorated soils, a dose of 1.6 to 2.4 tonnes per ha will be needed. Before the application of gypsum, lumps should be broken to at least coarse sand size (< 2 mm) and mixed with the soil in the shallower depth (10-20 cm). Deeper mixing results in diluting the effect of gypsum.

In highly deteriorated soils if it is intended to push the salt below 30 mm, a total of 60 cm water will be needed (Bhumbla, 1972). According to Abrol and Fireman (1977), the salinity level is reduced to about one-fifth its original value by pasage of unit depth of water per unit depth of soil. For example, to reduce the average salinity in the top 60 cm of a soil to one-fifth its original value, it would require passage of about 60 cm of water through the soil and to reduce the salt to 10 per cent of its original value would require about twice as much water. Seldom is passage of more than 100 cm of water through saline soil required for complete reclamation. Advantage cane be taken of rain water along with the application of rice husk as a mulch at the surface. This practice would reduce the upward movement of salt.

Leaching requirements: The leaching requirement is defined as the fraction of irrigation water that must be leached through the root zone to control soil salinity at any specified level. The amount of water required to leach a saline soil will be several times greater than the leaching requirement determined from the following formula to maintain a healthy soil.

$$LR = \frac{Dd}{D_i} \quad \frac{EC_i}{EC_d}$$

Where, LR is the leaching requirement, Dd and Di are the depths of drainage water and irrigation water, respectively, and ECi and ECd are the electrical conductivities of the irrigation and drainage waters.

Leaching requirement increases at least 10 percent for each 1000 parts per million (ppm) of dissolved salts in irrigation water. Hence, irrigation water having 4000 ppm will necessiate that approximately 50 per cent of applied water passes through the root zone

2) **Scraping of top soil affected with salt**: When the soluble salts accumulate on the surface as efflorescence, they may be removed to some extent by scraping off the surface soil. This is temporary cure and salinity again develops on such lands.

2. Agronomic techniques for management

It is not always possible to get rid of all the excess salinity. Under such situation, soils can be cropped with some adjustment in production by adopting special management practices, like selection of salt tolerant crops and varieties, planting method so as to minimise salt build up in the immediate vicinity of the germinating seed or plant root, and irrigating more often (Srivastava et al., 1979).

1) **Choice of Crops and Cropping Systems:** Salt tolerance of a given species depends upon three attributes (Hayward and Wadleigh, 1951)
    a) the capacity to increase the osomatic pressure of the tissue fluids to compensate for increases in osomotic pressure of the substrate,
    b) the capacity to regulate the intake of ions so as to bring about the increase in osomatic pressure and yet avoid an excess accumulation of ions, and
    c) The inherent ability of the protoplasm to resist deleterious effects of accumulated ions.

It is apparent that the main deficiencies of economic crops which lack salt tolerance are the inability to regulate adequately the intake of salt and the specific sensitivity of their protoplasm to accumulations of salt within the tissues.

Based on a large number of studies, the relative tolerance of crops to salinity is given in Table 8. Crop yields are seriously affected if the electrical conductivity of the saturation extract is more than about 12, 6 and 4 ds per m in respect of tolerant, semi-tolerant and sensitive crops, respectively. Tolerance in each column of Table 8 decreases from top to bottom.

TABLE 8: Relative Tolerance of Crops to Salinity (saline soils) (Abrol and Fireman, 1977)

Tolerant	Semi-tolerant	Sensitive
Date Palm	Pomogranate	Citrus
Barley	Wheat	Cowpeas
Sugarbeet	Oats	Gram
Spinach	Rice	Peas
Rape	Sorghum	Groundnut
Cotton	Maize	Guar
	Sunflower	Lentil
	Potatoes	Greengram

Another classification of various crops according to their tolerance is given in Table-9.

TABLE 9: Crops groups based on Response to salt stress (Mass & Hoffman, 1977)

Sensitive group		Resistant group	
Highly sensitive	**Medium Sensitive**	**Medium tolerant**	**Highly tolerant**
Lentil	Radish	Spinach	Barley
Mass	Cowpeas	Sugarcane	Rice (transplanted)
Chickpea	Broad-bean	Raya	Cotton
Beans	Vetch	Rice (direct sown)	Sugarbeet
Peas	Cabbage	Wheat	Turnip
Carrot	Cauliflower	Pearl millet	Tobacco
Onion	Cucumber	Oats	Safflower
Lemon	Gourds	Alfalfa	Taramira
Orange	Tomato	Blue panic grass	Karnal grass
Grapes	Sweet potato	Para grass	Date palm
Plum	Millets	Sudan grass	Mesquite
Pear	Maize	Guava	Tamarix
Apple	Clover	Pomegranate	Salvadora
	Berseem	*Desi babul (Acacia)*	

Earlier, Uppal *et al.,* (1961) prepared a list of crops that can be grown at various stages of reclamation and can be categorized as high, medium and low salt tolerant crops:

i) High salt-tolerant crops

Kharif (Autumn): Dhaincha (Sesabania aculeata)

Rabi (Winter): Oats, berseem, lucerne, senji (Trigonell sp.), methi (Fenugreek) and barley.

ii) Medium salt tolerant crops

Kharif (Autumn): Castor, cotton, sorghum, pearl millet and maize

Rabi (Winter): Mustard and wheat.

iii) Low salt tolerant crops

Kharif (Autumn): Greengram, blackgram, pigeonpea and sunnhemp.

Rabi (Winter): Chickpea, peas, linseed, sesamum

As far as possible crops are to be taken in the order of given above.

Certain varieties of crops are more salt tolerant than others. Varietal differences in such crops are reported to be of little practical significance.

In rice, coarser varieties have higher tolerance than finner varieties. Late rice is more salt tolerant than early rice.

In arid and semi-arid regions of north-western India, cotton, sorghum, pearl millet, clusterbean and mothbean should be grown during *Kharif*, and wheat, barley, mustard and safflower during *rabi* seasons depending upon the availability of water (Singh *et. al,* 1998). The threshold values for wheat, barley and mustard were found to be 4, 7 and 6 ds per m, respectively (Sharma *et.al*, 1990). Singh and Sharma (1993) reported that pearlmillet and cotton can be grown in saline soils up to ECe value of 6.5 and 7.5, respectively without significant reduction in yield.

For water deficient areas, crops with low consumptive use (evapotranspiration) or high tolerance should be selected. The recommended and remunerative cropping sequences for saline soils are pearlmillet-barley/wheat/mustard, and sorghum (fodder)-wheat/barley/mustard.

In water deficient areas mustard could replace wheat since its water requirement is low as compared to wheat.

Most of the coastal saline soils of the eastern regions are monocropped with rice during *kharif* season because no other crop survive owing to excess water conditions. Wherever irrigation water is available barley, linseed, mustard, safflower, chilli and coix (fooder) may be grown during *rabi* season with use of mulches. Wherever ample water for irrigation is available rice-rice and rice-rice-*dhaincha* (sesbania) showed more scope in terms of grain yield and soil improvement.

Vegetable crops: The high salt tolerant vegetable crops are garden beet, asparagus and spinach. The medium salt tolerant vegetables are tomato, Broccoli, cabbage, cauliflower, lettuce, sweet corn, carrot, onion, peas, squash and cucumber. The salt sensitive vegetables are radish, celery and green beans (US Salinity laboratory staff, 1954).

Fruit crops: With a few exception, most of the fruit trees are relatively sensitive to salinity and prone to leaf injury if sodium and chlorine are accumulated in soils. Peas, apple, orange, apricot, strawberry, lemon, avocado and papaya are sensitive even to 2-3 ds per m conductivity of soil saturation extract. Date palm is high salt tolerant, while pomogranate, fig and grape is medium tolerant (US Salinity Laboratory Staff, 1954).

According to Raheja *et.al* (1962), of the sensitive citrus crops the order of tolerance is lemon > sweet lime > Nagpur organe > mosambi, guava, mango and jackfruit are relatively more salt tolerant.

Grasses, legumes and trees: Singh (1990) recommended that the utilization of these problematic salt affected lands can be considered by growing salt tolerant suitable species as given below:

Grasses:	*Cynodon dactylon, C. plectostachyon, Paspalum notatum, P. dilatatum, chloris gayana, Lasiurus sindicus, Brachiaria mutica, Sporobolus marginatus, Urochloa mossambicensis.*
Legumes:	*Glycine javanica, Macroptilum* sp., *Phaseolus junatea, Stylosanthes humilis, Lotonis bainesil*
Tress	*Acacia tortilis, A. nilotica, P. juliflora, Dendrocalamus strictus, Salvadora* sp., *Zizyphus* sp., *Sesbania* sp., *Albizia amara*

2) **Nutrient Management:** Excess soluble salts in saline soils have a confounding influence on mineral nutrition of plants. Contrary to alkali soils, the root environment in saline soils is saddled with excess of neutral soluble salts mostly Cl^- and SO_4^{--} of N^+, Ca^{++} and Mg^{++} and pH of the saturated soil paste is less than 8.2. Excessive presence of salts cause growth reduction or failure in plants through high osomotic pressure of soil solution causing low physiological availability of water, direct toxic effect of individual ions and a complex interaction between Na, Ca and K resulting in the disturbed equilibrium of these ions in the plant. At the result of this, barley plants show increased uptake of Na, Mg, Mn and Zn but decrease uptake of P, K, Ca, Cu and Fe.

For most crop plant, yields decrease with increase in salinity but for a given salinity level, the yields can be increased by judicious application of fertilizers. However, response to added fertilizer are limited to low to medium level of salinity, normally between 8 to 10 ds per m.

a) **Nitrogen**: Saline soils are inherently deficient in nitrogen. Symbiotic N-fixation is also poor due to toxic effect of salts on rhizobia. The higher concentration of salts (Kcl and K_2SO_4) inhibit nitrification resulting in reduced production of NO_3^{--} - N and accumulation of NH_4^+ - N. Hence, it is better to use nitrate than ammonical fertilizer in split application. The N requirement of crops is higher in saline soils than in normal soils. Nitrogenous fertilizer should be applied in split doses to reduce N losses through volatilization and denitrification.

b) **Phosphorous**: Phosphorous application, generally, increases crop yields in saline soils. The availability of P in soil may be modified by soil salinity due to precipitation of a part of applied nutrient. The uptakes of P which is a relatively immobile nutrient in the soil is restricted because of poor root growth. At high salinity, the uptake of P by the plant may be severely affected due to competitive inhibitation caused by chloride as both P and Cl anions are absorbed by the same mechanism. However, the application of phosphatic fertilizers in saline soils helps in increasing the crop yields by directly providing phosphorous and by decreasing the absorption of toxic element like, chloride.

c) **Potassium**: Saline soils are generally medium to high in available potassium but plants grown under high salinity may show K deficiency due to antagonistic effect of Na and Ca on K absorption and/or disturbed Na/K ratio. Under such conditions, application of potassium fertilizers may give increased plant yields.

d) **Micro-nutrients**: Report of micronutrient deficiency in saline soils are uncommon. However, there are reports about decrease in the micronutrient content of plants with an increase in salinity.

Organic manures: The bulky organic manures have very high water holding capacity about 2 to 3 times its own weight. When sufficient amount of organic manures are added to saline soil, its water holding capacity increases resulting in decrease in conductivity of the soil saturation.

3) **Water Management:** In order to avoid development of salinity due to faulty system of irrigation, judicious use of quality water with low salt content at critical stages of plant, like germination / stand establishment and flowering is essential.

a) **Irrigation Scheduling**: The factors responsible for depth and frequency of irrigation are atmospheric evaporation, proliferation and depth of root penetration, capacity of soil to store and transmit water and the nature of plant response to soil water stress, etc. In saline soils, the osomotic pressure of the soil solution increases with the increase in the salt concentration and the plants are unable to extract water as readily as they can from a relatively normal soil. The crops tend to have poor crop stand and transpire less water. Hence, for keeping osomotic potential at low level, light and frequent irrigations should be applied to keep the soil moisture at field capacity. A heavy pre-sowing irrigation to leach the accumulated salts from the root zone is very useful and essential to improve germination and early growth.

b) **Quality of Irrigation water**: The standards for judging the quality of irrigation water as fixed in other countries are not applicable in monsoon type of climate. The new standard takes into consideration not only the water but also the soil and crop to be grown. Whereas, in heavy textured soil for semi-tolerant crop, the maximum permissible limit of electrical conductivity of water is 1500 micromhos, in light textured soil a water with an electrical conductivity of 8000 could be used for tolerant crops (Bhumbla, 1972)

TABLE 10: Standard for Irrigation Water

Water class	Electrical conductivity EC x 10^6	Salt Content		Sodium %	Boron ppm	Suitability
		Total ppm	Tons per acre - foot			
1	0 - 1000	0 - 700	1	60	0.0- 0.5	Excellent to good suitable for most plants under most condition
2	1000-3000	700- 2000	1 - 3	60-75	0.5- 2.0	Good to injurious, harmful to the sensitive crops
3	Over 3000	Over 2000	Over 3	75	Over 2.0	Unsuitable under most condition

The standard of water presented in Table 10 can be guide in appraising and developing irrigation and drainage practices to harmonise with qualities of water (U.S. Salinity Laboratory Staff, 1958).

c) **Economizing of water**: Substantial saving in irrigation water could be achieved by proper scheduling of irrigation as given below:

   i) Saline /drainage water may be used for raising salt tolerant crops in one season and good quality of water for salt sensitive crops in the second season.

   ii) Raising *rabi* (winter) crops with saline / drainage water and allowing other half of the year (normally monsoon period) for desalinization through rains (Singh and Chhabra, 1990).

   According to Sharma and Rao (1996), cyclic use of canal and saline waters is more beneficial to yield of crops and salinity build up than blending of these two waters. They further reported that the saline drainage affluent up to 9-10 ds per m can be utilized successfully for winter crops.

d) **Improvement of drainage**: Salinity developed due to restricted drainage may be improved by breaking of hard pans and providing adequate drainage.

   i) **Adequate Drainage**: Drainage ensures the removal of water that may be ponded over the land surface and removal of excess water from the potential crop root zone below the ground surface. In salty soils with high water table, artificial drainage in necessary before the excess salts can be removed. This can be achieved by (a) construction of drainage canals, (b) under drainage, and (c) pumping the water to keep the water table low when impeded drainage due to high water table.

   Sub-surface drainage provides an efficient means for lowering water table and to prevent secondary salinization. In practice, leaching of salts below the root zone and not allowing them to come up are difficult to accomplish if the soils are deep and heavier in texture, particularly if lower layers contain more salts than the surface layer.

Disposal of drainage water assumes serious problem in areas where no natural outlet, like sea is available. The problem can be solved to some extent by (a) minimising leaching and thus, the volume of drainage water, (b) disposal of saline drainage water by evaporation in salt pans and collecting drain water in selected areas for later injecting into appropriate deep aquifer or temporary surface storage let off during rainy season making use of heavy runoff to dilute the drainage effluent, (c) using stored drainage water for brackish water fish culture, and (d) use of such water in conjunction with good quality water for irrigation.

ii) **Breaking of hard pans**: Deep chiseling or deep ploughing is sometimes used on soils with imprevious layers in order to open the soil for the desired downward movement of water and salt.

Cultural Practices

i) **Bunding and land levelling**: Satisfactory land levelling is most important step in any reclamation process. To achieve this, the field should be provided with 35 to 40 cm high strong bunds all along the field boundary. The whole field should be divided into sub-plots of 0.4 ha size. Bunding helps in retaining the entire rainwater and in preventing the outside water from entering the fields. A gentle slope of about 0.1 per cent towards the drainage channel helps in taking out excess water when required. Proper land levelling helps in proper water distribution and leaching of the soluble salts.

ii) **Planing seeds on the side of ridges or in furrows**: One of the most serious problems on saline soils is the partial or complete failures of seed germination, poor crop stand and stunted plant growth resulting in low or no harvest. Water generally evaporate from the top surface by capillary and hence, top surface has maximum salt concentration. If the seeds are planted at the sides of the ridges or inside the furrows, they escape the zone of maximum salt concentration and thus, can germinate and develop properly during their early stages. Later on, when the roots are sufficiently developed and go deep down, the position of the ridges can be altered. It has been reported that trench method of planting sugarcane gave significantly higher yield than flat sowing, which may be attributed to less concentration of salts, which move to the top of adjoining areas (Dargon et al., 1973). Similarly, in case of sugarbeet it was observed (Dargan and Chillar, 1973) that sowing of sugarbeet on the side of the ridge half way between top and bottom, opposite to the sun direction gave significantly higher yield than flat and ridge sowing. The salt concentration in the root zone under this method of sowing was comparatively less owing to greater accumulation of salts in southern face of ridge due to direct and intensive solar radiation on this face.

iii) **Alternate- ditch - irrigation**: The investigation carried out at IARI indicates that it is possible to grow salt sensitive crops, such as maize, sorghum, cowpeas, tomato, potato, cotton, mustard, cabbage, cauliflower, radish, carrot, etc. by a simple and inexpensive agro-technique known as the 'alternate-ditch-irrigation' on saline soils. This technique consists in preparation of ridge seedbeds on saline soils, irrigation of alternate ditches, the ridges of which are sown with crops, thus leaving every other ditch unirrigated and unsown. This system of irrigation flanks the salt to the sides of the unirrigated ditch and thus, desalnising the irrigated slopes and rendering them safe for seed germination and establishment of transplants.

Alternate-ditch-irrigation of ridge seedbeds affects economy in water use to the extent of about 50 per cent as only the half area is irrigated, operates irrespective of the ridge direction and above all ensures good germination stand, growth and yield of crops. A ridge size of 22.5 to 30 cm in height and ridges spaced 60 cm apart have been observed to suit very well to most of the field and vegetable crops. The unirrigated ditches can be made little narrower and slightly deeper than the irrigated ditches so as to put more area under the crops and enhance the efficiency of unirrigated ditches as surface drains in monsoon. The ridges can be made with an inversion plough or a suitable ridger. This technique of alternate-ditch-irrigation can also be extended to alkali soils after converting these soils into saline soils by use of an amendment like gypsum.

iv) **Construction of wind-break to restrict wind blown salt**: When salinity is developed due to wind blown salt, construction of wind-break along with other practices, like leaching, judicious application of irrigation water with adequate drainage should be followed.

v) **Construction of bunds to prevent tidal water**: Coastal saline soils are subjected to tidal flooding. The amptitude of tides in India varies from 0 to 6 m (Rao and Dhruvanarayan, 1978). Accordingly, bunds should be constructed to check the flow of tidal waves. Introduction of artificial drainage is also essential.

vi) **Maintenance of Plant Population**: Mortality of young seedlings and poor tillering of the crops are common in saline soils. Higher seed rate and closer spacing are recommended to counter these effects. For cotton, pearlmillet, sorghum, wheat, barley, mustard etc. about 25 per cent higher seed rate over that recommended for normal soils ensures good crop stand. In coastal saline soils, 3-4 seedlings per hill are required in *Kharif* season to compensate the yield loss in rice due to seedling mortality. In case of high yielding varieties, 15 x 15 cm spacing was found to be optimum for maximum grain yield.

vii) **Seed treatment**: The chemical treatment of seeds and seedlings has been reported to induce salt tolerance. Seed treatment with calcium salt increased the salt tolerance of wheat in saline soils. In a field experiment on saline soils (EC_e12ds/m), irrigated with saline water (EC_{iw} 10.5 ds/m), the highest yield of wheat was obtained when seeds soaked in 3 per cent sodium sulphate solution were used (Puntamkar et al., 1971)

B. ALKALI SOILS

Alkali soils, also known as sodic soils contain sufficient exchangeable sodium to interfare with the growth of most crop plants, either with without appreciable quantities of soluble salts.

Alkali soils differ from salins soils in terms of high pH and ESP, high content of calcium carbonate, presence of hard calcic (*kankar*) layer, low infiltration and presence of shallow cracks in the surface layer on drying. While saline soils mostly have brackish underground water, alkali soils normally have good quality or sodic underground water. The word 'alkali' is of Arabic origin meaning ash - like. Russian terms them 'solonet'.

Alkali soils have been divided into three types depending on their chemical characteristics:

a) **Saline-Alkali Soils**: The term 'saline-alkali' applies to soils which are both saline and alkali. There can be all stages in transition with varying degree of dominance of salt content or high pH.
b) **Non-Saline-Alkali**: These soils have often be called 'black alkali' soils, because they are black owing to the effect of the high sodium content which causes the dispersion of the organic matter.
c) **Degraded Alkali**: Such degraded soils are called 'solodi' or 'solod' by Russian workers and the process of formation is termed 'solodization'.

Characteristics

Chemical characteristics

(i) (a) Saline-Alkali soils are characterised by ESP > 15, EC 4 ds per m and a variable pH of saturated soils paste, usually above

8.2 depending upon the relative amounts of exchangeable sodium and soluble salts. The ratio between soluble Na and Cl + SO_4 is always less than one. When soluble salts are leached downward, the pH will rise above 8.2, but when soluble salts again accumulate, the pH will again fall to 8.2. Physical condition is that of saline soils i.e. good permeability and good aggregation but contrary to saline soils become dispersed on leaching and removal of soluble salts and they assume all the characteristics of non-saline alkali soils including increase in pH above 8.2.

In presence of free bicarbonates and carbonates the availability of Fe, Zn, Cu and P decreases. When the soluble salts also contain zypsum the physical deterioration may be avoided.

(b) Non - Saline - Alkali soils are those in which the percentage of exchangeable sodium saturation is > 15, as a result, the pH is between 8.2 and 10.0. The conductivity is < 4 ds per m.

Anions present are chloride, sulphate and bicaronates. Some amount of carbonate is always present. Na is predominant cation and Ca and Mg occupy a small portion of the adsorbed cations. Potassium may be present in considerable amount.

Soils are highly dispersed and have a very poor physical condition. Clays may be badly lost from the surface and deposited in the sub soil forming hard pan. The surface soils may become coarse textured and fribale.

The availability of P, Ca, Zn, Mo and Fe decreases, sodium may become toxic and nitrificaton may be slowed down.

Degraded Alkali soils are those soils which have adsorbed sodium more than 15 per cent but the exchangeable H^+ ions, in absence of a very low amount of calcium and magnesium, is so high as to give an acidic pH of surface. The pH of the subsoil, where adsorbed Na ions predominate over other cations, is above 8.2.

Physical properties are like that of non-saline-alkali soils, but there is improvement in structure of the portion of soil where the pH is acidic.

(ii) General availability of nutrients in Alkali soils

- Phosphorus, Cu, Zn, Mo, and Fe become less available in presence of free carbonate ions as these are precipitated to various insoluble forms.
- Alkali soils, being poor in organic matter, contain very low amounts of available N.
- Gypsum is always absent in such soils but insoluble calcium carbonate is always found in variable amounts.
- Ammonification and nitrification may proceed but conversion of nitrite to nitrate is reduced and hence, there may be accumulation of nitrites. There is also chances of loss of N in ammonical form under high pH and alkaline condition.
- High available Na may introduce Na-toxicity, and also reduce the absorption of Ca and Mg by plant roots. Soil may also suffer from B and Mo toxicity.

Physical Characteristics

a) Excess exchangeable sodium and high pH result in dispersion of clay particles which in turn block the soil pores. These soils have an unstable structure which is not conductive to free transfer of air and water within the root zone. The hardening and crust formation of surface soils upon drying may often cause a great hinderance to the emerging seedlings.

b) Infiltration and permeability of alkali soils is very poor. These soils are compact and dense and offer high mechanical impedence to the growth of roots.

c) The presence of calcic (*Kankar*) horizon of 30 to 60 cm width around one meter depth in the profile is a severe physical impediment for downward growth of roots and capillary movement of moisture within the profile.

d) Alkali soils permit excessive runoff losses as moisture retention capacity of such soils is very poor. Runoff leads to soil erosion and loss of nutrients.

Microbiological activity

i) When pH is high (> 9.0) and the total nitrogen content of soil low (<0.2%), the population of algae flora is poor.
ii) At the lower pH of 7.9 - 8.6, the population of blue green algae is increased in stagnant water and ditches.

Development of Alkali Soils

As in the case of saline soils, alkali soils are also generally limited to arid and semi-arid climates, where the evaporation is more than precipitation or where the drainage is impeded. They are also occasionally found in semi-humid regions of restricted drainage, or where the areas previously getting salts due to tidal waves are subjected to leaching.

Source of Sodium salts: The sources of sodium salts may be due to:

i) weathering of rocks and sodium bearing minerals,
ii) washing from the surrounding uplands,
iii) salt blown from the sea,
iv) tidal waves,
v) irrigation water, and
vi) excess irrigation which links the percolating water to the already existing saline ground water which comes up by capillary action during dry months.

Stages of evolution: Alkali soil may pass through five stages in their evolution and development:

i) *Salinization* - The process of accumulation of salts in the soil.
ii) *Alkalization* - A process whereby the exchangeable sodium percentage in the soil is increased.
iii) *Desalinization* - The process of leaching out of soluble salts.
iv) *Degradation* - The process of replacement of adsorbed Na^+ ions by exchangeable H^+ ions.

v) *Regradation* - The process of reoccurence of salts.

Process of development: The development of three different kinds of alkali soils is given below separately:

Saline - Alkali Soils

In those places where soluble salts contain sodium more than 50 per cent of the total soluble cations as considerable portion (more than 15%) of the cation exchange capacity is fulfilled by sodium, the soil becomes saline-alkali soil.

The same factors responsible for development of saline soils are also responsible for the development of saline-alkali soils. Accumulation of salts can affect any group of soils where conditions become favourable for salt accumulation, i.e., absence of leaching, restricted drainage etc. With the salinization of profile, a series of reactions take place concurrently. If the soluble cations consists of predominantly sodium salts, considerable Ca, Mg and K are displaced from the exchange position of the clay complex and Na gets adsorbed as a result of base exchange. The process is known as alkalization and the soil passes to saline-alkali stage.

Non-Saline-Alkali Soils

These soils develop in two ways: (i) by removal or washing off the soluble salts of a saline-alkali soil (ii) direct development without going through the saline-alkali stage.

In the saline-alkali stage, movement of salt takes place either to the surface by evaporation or downward by leaching. Introduction of drainage or irrigation is also instrumental in washing away of soluble salts. The soil contains more than 50 per cent of sodium as soluble cations, and hence, have a higher molar concentration in the soil solution. More sodium is adsorbed on the clay surface than other cations. The adsorbed sodium rise considerably over 15 per cent and the pH increases. Removal of excess salts brings about a process of desalinization accompanied by significant change in the constitution of soil profile both in the presence and absence of $CaCO_3$.

In the presence of free $CaCO_3$ exchange between the Na+ of clay and Ca^{++} from the weakly ionized $CaCO_3$ may take place.

$$\text{Na - clay} + H_2O = \text{H - clay} + NaOH$$
$$\text{2H - clay} + 2\,NaOH + CaCO_3 = \text{Ca - clay} + Na_2CO_3 + 2H_2O$$

In the presence of free $CaCO_3$, some of the exchangeable Na is hydrolysed giving rise to NaOH which combines with CO_2 of soil air and forms Na_2CO_3.

$$\text{Na - clay} + H_2O = \text{H - clay} + NaOH$$
$$2\,NaOH + CO_2 = Na_2CO_3 + H_2O$$

This Na_2CO_3 imparts a very high pH to the soil which may be as high as 10 in some cases. This Na_2CO_3 may even react with any Ca-clay that comes in contact converting it into Na-Clay and precipitating $CaCO_3$. Either way the soil gets strongly alkaline. Hydrolysis may even bring about formation of slica gels. These together with humates may work as a cementing agents in producing a hard pan in the sub-soil.

When soluble salts high in sodium cations are being washed down to the basin shaped land, the soluble salts go down to the centre of the basin and accumulate there forming saline-alkali or saline soil. At the margin and lower slopes of the basin, soluble salts do not accumulate, but these areas are also in contact with solution containing higher concentration of sodium. Thus, sodium is adsorbed on this area replacing other cations and a non-saline-alkali soil develops. The development of non-saline-alkali soil is facilitated in the areas where saline-alkali soils are subjected to leaching either by change in climate or introduction of drainage and irrigation.

Degraded Alkali Soils

When a non-saline- alkali soil, which has practically no $CaCO_3$, is subjected to further leaching, the hydrogen ions of the water replaces sodium ions adsorbed on the clay surface. As a non-saline-alkali soil has got very low permeability, the drainage is mostly lateral and the replacement of sodium by hydrogen ions is in horizontal sheets resulting development of acidic condition on surface soil, whereas the soil just below is highly alkaline. The acidic layer gradually extends downwards and with its downwards movement it starts to destroy the hard pan and columnar structure of the subsoil.

Plant Growth on Alkali Soils

Alkali soils may seriously inhibit or entirely prevent plant growth, yield and/ or quality by anyone of the following attributes:

i) the availability of Ca, Mg and micro-nutrients like Zn, Fe and Mn effectively depressed by the relative high percentage of adsorbed alkali cations, particularly Na on the exchange complex of these soils. The decreased solubility and availability of micronutrients may be due to high pH, insoluble Ca_2CO_3 and soluble CO_3^{2-} and HCO_3 ions,

ii) the activity of the hydroxyl and carbonate ions may be sufficiently high to be toxic to the plants. Excess of Na too may be toxic to the plants causing imbalance due to antagonestic effect on K and Ca nutrition. Increased solubility and accumulation of elements, like flourine (F), selenium (Se) and molybdenum (Mo) in toxic levels in plants may effect not only crop yield but also health of animals feeding on them, and

iii) an accumulation of adsorbed Na on the exchange complex may have a dispersive effect on soil aggregates resulting poor water and air permeability, and thereby, bringing about puddled condition, low availability of water due to poor conductance from lower to upper soil layers, and hard crust on the surface layer reducing germination and seedling emergence.

For most crop plants, the calcium becomes unavailable when the exchangeable Na percentage approaches 50. Few crops can thrive under alkalinity. The low level of CO_2 may be a major limiting factor in the lack of availability of phosphate in alkaline - calcareous soil.

Alkali soils readily become dispersed and a dispersed soil is not conductive to vigorous growth due to poor tilth. They get pudled easily. The puddled soil has lower capacity for gaseous exchange which may result in oxygen deficiency at the absorbing surface of the roots and most plants are unable to take in adequate quantities of water and nutrients at low oxygen stress.

The very poor structural status of the alkali soils in the barren areas results in the poor infiltration causing the soil to be deficient in moisture most of the time.

Management of Alkali Soils

Management of alkali soils involves their reclamation by soil amendments and efficient agronomic practices for sustainable crop production.

Reclamation of Alkali Soils

The reclamation methods for different kind of alkali soils are different but the principles involved are the same as follows:

A. Removal of soluble salts if present.
B. Replacement of adsorbed sodium ions by calcium ions.
C. Removal of causes that are responsible for the production of alkalinity, e.g., high water table, impeded drainage, high sodium absorption ratio for irrigation water, etc.

A. **Removal of soluble salts**: This is only in the case of saline-alkali soils. The methods applied are the same as discussed in the case of saline soils but in case of soils which have no $CaSO_4$ present as soluble salt, they develop into non-saline alkali soils on removal of excess salts. Hence, care should be taken to replace adsorbed sodium by calcium ions by application of gypsum, etc. in the water used for flooding or irrigation.

B. **Replacement of adsorbed sodium ions using amendments:** Amendments are the substances that influence the plant growth favourably by changing the soil reaction favourably, making plant nutrients available, improving the physical condition of soil and countering the effect of injurious substances. The principal purpose of amendments is to replace exchangeable sodium or to neutralize alkaline salts with acid. In the present case, for reducing exchangeable sodium percentage (ESP) of the soil for crop production, amendments are used for directly supplying calcium to replace exchangeable sodium or supply calcium indirectly by dissolving calcite (natively found in alkali soils) because of their acidulation effect or by dissolving native calcium carbonate by increasing partial pressure of carbon-di-oxide. The calcium so mobilised is used to replace sodium from the exchange complex and thus, reclaiming alkali soils.

Chemical amendments: Chemical soils amendments used either to add calcium or hydrogen ions are of the following three types:

i) Soluble calcium salts
   a) Calcium chloride
   b) Calcium sulphate (gypsum)

ii) Acid or acid formers
   a) Sulphur
   b) Sulphuric acid
   c) Iron sulphate
   d) Aluminium sulphate
   e) Lime sulphur (CaS_5)

iii) Calcium salts of low solubility
   a) Ground limestone
   b) Byproduct lime from sugar factories

While each type of amendments has a place in alkali reclamation, proper use and effectiveness of these chemical amendments under different soil conditions are governed by several factors, the principal ones being the alkaline earth carbonate content and pH reading. The gypsum equivalent of different amendments for supplying calcium in terms of pure gypsum (100%) is presented in Table 10.

TABLE 10: Gypsum Equivalent of Different Amendments

Amendment	Quantity (tonnes)
Gypsum ($CaSO_4$, $2H_2O$)	1.00
Calcium chloride ($Cacl_2$, $2H_2O$)	0.85
Calcium nitrite [Ca $(NO_3)_2$, $2H_2O$]	1.06
Sulphur (S)	0.19
Sulphuric acid (H_2SO_4)	0.57
Pressmud (Lime-sulphur, 9% Ca, 24% S)	0.78
Iron sulphate ($FeSO_4$, $7H_2O$)	1.62
Aluminium sulphate [$Al_2(SO_4)_3$, $18H_2O$]	1.29
Iron pyrite (FeS_2, 30% S)	0.63

(Source: Based on Abrol and Fireman, 1977)
Note: Agricultural grade gypsum is only 75 to 80 per cent pure. In case of sulphur, effectiveness is much lower than other amendments since in practice 100 per cent oxidation does not happen as assumed.

The chemical reaction of each amendment along with kind of alkali soils on which they should be used is given below:

Saline - Alkali Soils

1) **Gypsum**: It is a natural sulphate of lime (calcium) found in large deposits in Rajasthan. In saline-alkali soils, gypsum is applied with flooding water or irrigation water. It reacts with exchangeable sodium which gets converted into sodium sulphate which is leached from the soil to reduce the soil pH. The following chemical reaction takes place:

$$\underset{\text{(Alkali Soil)}}{2Na\text{ - Soil}} + \underset{\text{(Gypsum)}}{CaSO_4} \longrightarrow \underset{\text{(Normal Soil)}}{Ca\text{ - Soil}} + \underset{\text{(Sodium sulphate)}}{Na_2SO_4}\downarrow$$

2) **Sulphur** (also sulphuric acid): Sulphur is a very effective to replace exchangeable sodium. Theoretically one atom of S replaces four Na ions by calcium. But under field conditions, the actual replacement of exchangeable Na is much lower. The chain reactions take place with addition of sulphur is given below:

$$\underset{\text{(Sulphur)}}{2S} + \underset{\text{(Oxygen)}}{3O_2} \longrightarrow \underset{\text{(Sulphur Trioxide)}}{2SO_3} \quad \text{(Microbial Oxidation)}$$

$$\underset{\text{(Sulphur trioxide)}}{SO_3} + \underset{\text{(Water)}}{H_2O} \longrightarrow \underset{\text{(Sulphuric acid)}}{H_2SO_4}$$

$$\underset{\text{(Sulphuric Acid)}}{H_2SO_4} + \underset{\text{(Calcium carbonate)}}{CaCO_3} \longrightarrow \underset{\text{(Calcium sulphate)}}{CaSO_4} + \underset{\text{(Carbon dioxide)}}{CO_2} + \underset{\text{(Water)}}{H_2O}$$

Or

$$H_2SO_4 + 2CaCO_3 \longrightarrow CaSO_4 + \underset{\text{(Calcium bicarbonate)}}{Ca(HCO_3)_2}$$

$$\underset{\text{(Alkali Soil)}}{2Na\text{ - Soil}} + \underset{\text{(Calcium Sulphate)}}{CaSO_4} \longrightarrow \underset{\text{(Normal Soil)}}{Ca\text{ - Soil}} + \underset{\text{(Sodium sulphate)}}{Na_2SO_4} \downarrow$$

3) **Lime sulphur**: When like sulphur is applied to saline-alkali soil, following reaction takes place:

$$\underset{\text{(Lime sulphur)}}{CaS_5} + \underset{\text{(Oxygen)}}{8O_2} + \underset{\text{(Water)}}{4H_2O} \longrightarrow \underset{\text{(Calcium sulphate)}}{CaSO_4} + \underset{\text{(Sulphuric acid)}}{4H_2SO_4}$$

$$CaCO_3 + H_2SO_4 \longrightarrow CaSO_4 + CO_2 + H_2O$$

$$\underset{\text{(Alkali Soil)}}{2Na\text{ - Soil}} + \underset{\text{(Calcium Sulphate)}}{CaSO_4} \longrightarrow \underset{\text{(Normal Soil)}}{Ca\text{ - Soil}} + \underset{\text{(Sodium sulphate)}}{Na_2SO_4} \downarrow$$

Sodium sulphate being a soluble neutral salt is lost in drainage.

4) **Iron sulphate** (also aluminum sulphate) / Pyrites

$$\underset{\text{(Pyrites)}}{2FeS_2} + 2H_2O + 7O_2 \longrightarrow \underset{\text{(Iron Sulphate)}}{2FeSO_4} + 2H_2SO_4$$

$$H_2SO_4 + CaCO_3 \longrightarrow CaSO_4 + CO_2 + H_2O$$

$$2Na\text{ - Soil} + CaSO_4 \longrightarrow Ca\text{ - Soil} + Na_2SO_4 \downarrow$$

Non-Saline- Alkali Soils

i) **Gypsum**: Reaction as given above for saline-alkali soils.
ii) **Sulphur**: Reaction as for saline-alkali soils.
iii) **Sulphuric acid**: Reaction takes place as given below:

$$2S + 3O_2 \longrightarrow 2SO_3 \text{ (Microbiological oxidation)}$$

$$SO_3 + H_2O \longrightarrow H_2SO_4$$

$$\underset{\text{(Alkali soil)}}{2Na\text{ - soil}} + H_2SO_4 \longrightarrow H\text{ - soil} + Na_2SO_4 \downarrow$$

iv) **Lime sulphur**: Reaction of lime sulphur in non-saline-alkali soil is given below:

$$CaS_5 + 8O_2 + 4H_2O \longrightarrow CaSO_4 + 4H_2SO_4$$

$$\underset{\text{(Alkali soil)}}{Na\text{ - Soil}} + CaSO_4 \longrightarrow \underset{\text{(Normal soil)}}{Ca\text{ - Soil}} + Na_2SO_4 \downarrow$$

v) **Iron sulphate**: Reaction of iron sulphate in non-saline-alkali soil takes place as follow:

$$FeSO_4 + H_2O \longrightarrow H_2SO_4 + FeO$$

$$\underset{\text{(Alkali Soil)}}{Na\text{ - Soil}} + H_2SO_4 \longrightarrow H\text{ - Soil} + NaSO_4 \downarrow$$

Degraded Alkali Soils

i) **Gypsum**: Same as in case of saline-alkali soils
ii) **Sulphur**: Same as in case of non-saline -alkali soils
iii) **Iron sulphate**: Same as in case of non-saline-alkali soils
iv) **Limestone**: This amendment is essential for degraded alkali as the top soils become acidic

$$\underset{\text{(Degraded alkali)}}{\text{H - Soil}} + CaCO_3 \longrightarrow \underset{\text{(Normal soil)}}{\text{Ca - Soil}} + CO_2 + H_2O$$

When limestone is applied to the soil, it gets dissolved into the soil solution. The calcium part of the limestone reacts with the soil complex and replaces sodium. Sodium combines with carbonate cations to form sodium carbonate which is leached out of soil by flooding.

All chemical amendments are not suitable for all soil conditions. Gypsum is suitable for both saline and alkali soils, whereas sulphur and iron sulphate are suitable for non-saline-alkali and saline-alkali soils. These amendments are more effective up to pH of 9.0. Limestone is recommended in degraded alkali with acidic pH and in saline soils with pH less than 8.0. The alkali soils in Indo-Gangetic plains are mostly calcareous. For such soils, amendments, like calcium source (e.g., gypsum and phospho gypsum) or acid and acid formers are recommended.

Conversion of mineral sulphur to sulphuric acid is very slow in alkali soils since it takes place through biological oxidation by *Thiobacillus thioxidens*. Similarly, the biological oxidation of pyrites is carried out by *Thiobacillus ferroxidens* which are only active under low pH conditions. The sulphuric acid produced through oxidation reacts with calcium carbonate of the soil to release calcium for exchange with the sodium on the clay complex. Due to problems in oxidation, the efficiency of pyrites is 70 to 80 per cent as compared with gypsum at the recommended dose of application. Sulphuric acid as amendment does not go through the oxidation process and hence, reacts rapidly. However, sulphuric acid is highly corrosive and dangerous to handle.

Organic amendments: The decomposition of organic amendments releases CO_2 and other by-product, like acids or acidic products which enhance solubility of native $CaCO_3$ and thus, provide Ca for the removal of exchangeable Na. In addition to CO_2 byproducts of anaerobic decomposition of added organic matter make more of native calcium available and help in greater and faster removal of exchangeable sodium from the soils. On the other hand, aerobic decomposition does not increase appreciably the availability of calcium from calcium carbonate. Decomposition of green manure and other organic amendments, like straw, saw dust, rice husk, farm yard manure (FYM), compost, tree leaves, etc., under submerged condition can help to accelerate reclamation of alkali soils by producing carbon dioxide and organic acids, which help to increase electrolyte concentration, lower soil pH, and improve physical condition of soil. According to Swarup and Singh, (1993), the application of FYM at the rate of 20 t per ha when combined with gypsum gives higher yield than gypsum alone and hastens the reclamation of alkali soil. Organic materials together with inorganic amendments are more cost effective.

Amount of amendments: The amount of amendments needed to reclaim alkali soils depends on the soil pH and concentration of exchangeable sodium present in the soil complex. The amount of amendments should be such, so that to reduce sodium saturation to near about 10 per cent.

For everyone milli-equivalent (me) of Na to be replaced, one m.e. of Ca will be needed which amounts to 1.72 t gypsum for 0-15 cm depth of soil. Abrol and Bhumbla (1979) reported that 10 to 15 t of gypsum containing 70 per cent $CaSO_4$, $2H_2O$ which approximates about half of the actual gypsum requirement (GR) determined by Schoonover's method (Schoonover, 1952; Richard, 1954), is sufficient to reclaim the surface 15 cm to start with rice as first crop keeping in view the high tolerance of rice to soil exchangeable sodium percentage (ESP), its shallow root system and its capacity to solubilise soil $CaCO_3$ to further reduce soil ESP of surface and deeper layers over time so as to allow cultivation of other crops. Bhumbla (1970) found that for shallow rooted crops, application of gypsum at the rate of 1/4 of the gypsum requirement was adequate. However, in case of sensitive crops, like pulse high amounts of amendments right in the initial year of reclamaition are needed.

For the alkali soils of Indo-Gangetic plains of mostly calcareous nature, Central Soil Salinity Research Institute, Karnal has developed a relationship between pH of 1:2 soil water suspension and ESP and in turn with pH and gypsum requirement of the soil varying in texture and thus, cation exchange capacity. However, for each type of soil a relationship should be established separately (Abrol et al., 1980).

In the recent years, use of pyrites (FeS_2) has also been made on a large scale in some parts of the country.

Fineness of amendments: Gypsum lumps when ground to pass through 2-mm sieve contain sufficient fine sized particles to effectively reclaim alkali soils. Finner the gypsum particles more faster they react and also less amount of water is required for their dissolution (Chawla and Abrol, 1982). For maintenance of permeability of the alkali soils, a mixture of fine and medium sized particles may be preferred.

Frequency of application: Alkali soils once reclaimed do not need, generally, repeated application, provided these are under continuous croppings since calcium is added through fertilizers, like calcium ammonium nitrate and single super phosphate increasing the calcium carbonate content of soils. The added calcium, thus, replaces exchangeable sodium and also provides calcium need to crops. According to Chhabra and Kamra (2000) already reclaimed alkali soils may need repeated application when sodium is added from outside through the flooded or irrigation water, water table is shallow with high electrical conductivity, dry fallow period is prolonged, initial reclamation is incomplete because of low amendments or high water requiring crops are replaced by low water requiring crops resulting in upward flux of the reaction products.

Method of application: Unlike fertilizers gypsum is to be applied only once after first irrigation to ensure maximum leaching of soluble carbonate and nitrate so as to avoid its precipitation. The gypsum should be applied on soil surface and then mixed with disc or plough in the upper 6-8 cm soils two to four weeks before sowing. Water is ponded on the soil to ensure proper dissolution of gypsum and leaching of replaced sodium. Within 7 to 10 days of continuous ponding and leaching normally maximum amount (nearly 75 per cent) of applied gypsum gets dissolved.

Rice when taken as first crop, can be transplanted immediately but preferably after 2-3 days of gypsum application Post application leaching is not needed in rice because water used for growing rice can itself be used for dissolving amendments and to leach down the reaction product from the soil. However, in case of other crops taken as first crop, sufficient irrigation should be given to dissolve gypsum to leach down the reaction product and to reduce exchangeable sodium percentage.

Alkali soils contain usually high level of exchangeable sodium up to a depth of 100 cm or more, but it would suffice to improve the soil up to a depth of 15 to 20 cm by application of amendments to grow shallow rooted crops, like rice and wheat. For growing deep rooted crops, the quantity of amendments required would be much more.

C. **Removal of causes responsible for alkalinity development:** Alkali soils reclaimed by removal of soluble salts and by use of amendments may again develop alkalinity if the causes responsible for alkalinity development are not removed. Hence, to have a lasting effect care should also be taken of these factors as discussed below:

   i) **High water table**: This can be reduced by under drainage, drainage of basins and pumping out of water. Judicious and proper use of water may reduce water table developed due to faulty irrigation practices. High water table may also develop in light textured soil near canals with no lining. Lining of canals may reduce water table.

ii) **Impeded drainage**: This can be improved by providing proper drainage, under drainage, and pumping out of water. Impeded drainage developed as a result of hard pan can be improved by breaking or boring through the hard pans.

iii) **High sodium adsorption ratio of water**: This can be reduced by putting bags of gypsum in drainage canals. Lime-sulphur may also be used if available.

iv) **Increasing water holding capacity and buffering capacity of soil:** Addition of organic matter increases the water holding capacity and buffering capacity of soil which will reduce the harmful effect of soluble salts and adsorbed sodium ions, respectively. This can be done by applying bulky organic manures and green manure.

Agronomic techniques for management of alkali soils

(1) Choice of crops and cropping systems

The crops differ a great deal in their tolerance to alkalinity. Adverse conditions in alkali soils severely restrict the choice of crops to be grown. The crops that are more tolerant to high levels of exchangeable Na are the ones which normally take a considerable amounts of Na, whereas more sensitive crops are the ones which normally tend to exclude sodium. The tolerance of a crop to high percentage of adsorbed Na is modified by the pH of the soil and accumulation of soluble carbonate. Tolerance to soil alkali may involve the capacity by the plant to secure an adequate supply of calcium under conditions of relatively low availability. For most crop plants, the calcium becomes unavailable when the exchangeable-Na- percentage approaches 50.

The following classification gives the tolerance of crops to different levels of exchangeable sodium percentage (Branson, 1960).

TABLE 11: Tolerance of crops to Exchangeable sodium Percentage (ESP)

Classes	ESP	Crops
Very Sensitive	2 - 10	Deciduous fruits, nuts, citrus and avocado
Sensitive	10 - 20	Beans
Moderately tolerant	20 - 40	Clover, oats, tall fescue, rice and dallis grass
Tolerant	40 - 60	Wheat, cotton, alfala, barley, tomato and beats
Very tolerant	> 60	Crested and fairway wheat grass, Rhodes grass and tall wheat grass.

Relative tolerance of crops and grasses to exchangeable sodium percentage (Alkali soils) is presented in Table 12 (Abrol and Freeman, 1977).

TABLE 12: Relative Tolerance of Crops and Grasses to Exchangeable Sodium percentage (ESP)

Tolerant (ESP, 35-50)	Moderately tolerant (ESP, 15-35)	Sensitive (ESP, <15)
Bermuda Grass	Wheat	Cowpeas
Para Grass	Barley	Groundnut
Rice	Oats	Lentil
Sugar beet	Senji	Mash
	Berseem	Greengram
	Millets	Peas, Maize
	Cotton	Cotton (at germination)

Note: Tolerance in each column decreases from top to bottom.

In addition to the crops and grasses mentioned above legumes, like chickpea and pigeonpea are very sensitive to exchangeable sodium percentage (ESP) of less than 15. *Dhaincha* (*Sesbania aculeata*) - a green manuring crop is exceptional among legumes. Oilseeds except groundnut are generally moderately tolerant. Transplanted pearl millet perform better than direct sown crop.

Grasses like Karnal grass, (*Leptochloa fusca*) and Rhodes grass (*Chloris gayana*), Gatton panic (*Panicum maximum*), Bermuda grass (*Cynodon dactylon*) and para grass (*Brachiaria mutica*) are relatively more tolerant to alkali (Kumar and Abrol, 1986). The pH of the top 0-15 cm soil was reduced from 10.4 to 9.8 in a short span of one year by growing Karnal grass without amendment at a farmer's field where drainage was not feasible during rainy season (Kumar *et al.,* 1992). Growing of grasses in alkali soils results in progressive decrease in pH and ESP and increase in Ca content, infiltration rate and organic matter content in alkali soils.

During *Kharif* (rainy) season in the alkali soils, rice is favoured as first crop because it can tolerate standing water and medium levels of ESP. During *rabi* (winter) only shallow rooted crops, like wheat, barley, berseem and shaftal could be grown in the initial years, since only 0-15 cm surface soil depth is considered for reclamation of alkali soils. Therefore, rice based cropping sequences of rice-wheat/barley-sesbania and rice-berseem (Egyptian clover) / Persian clover are recommended for alkali soils during reclamation and for few years in the post-reclamation period. It is possible to adopt cropping pattern with relatively moderate and less tolerant and deep rooted crops, like sorghum / pearl millet, oilseeds, cotton and sugarcane. When pH of surface soils comes down to 8.5 to 9 after 3 years of crop sequence with rice, sorghum/ pearl-millet - wheat are suitable for areas with water scarcity (Singh and Chabbra, 1990).

On the basis of several agronomic trials, Dargan *et.al.* (1995) concluded that rice-wheat-dhancha (green manuring) and rice-berseem sequences were more efficacious and rumenerative. Singh et al. (1988) reported that oilseeds crop, like mustard, rapeseed and toria can also be grown as a first crop followed by sorghum and pearl millet for fodder in *Kharif (*rainy season*)* in a highly deteriorated alkali soils with the application of 50 per cent GR (0-15 cm depth) as gypsum.

(2) Cultural practices

Unlike normal soils deep ploughing and puddling in alkali soils are harmful practices. Tillering of rice and wheat in alkali soils is poor. To compensate for this poor tillering higher plant population and closer spacing are essential. In rice, planting 4 plants per hill instead of 2 results in increased rice yield. Similarly in case of wheat, seeding rate should be increased atleast by 50 per cent.

(3) Nutrient management

Availability of several essential nutrients is adversely affected due to high exchangeable Na, high pH, substantial presence of $CaCO_3$ content and high concentration of CO_3^{-2} and HCO_3 in alkali soils. Hence, the use of manures and fertilizers is very important in alkali soils.

Nitrogen

Alkali soils are deficient in nutrients particularly N. The soil is low in organic matter content and hence, poor in total and available N. Moreover, transformation of organic and inorganic form is not only slow but wasteful. A large fraction of applied N is also lost in gaseous form as ammonia under high soil pH conditions.

Of the nitrogenous fertilizers, ammonium sulphate is better fertilizer than urea or calcium ammonium nitrate. Complete hydrolysis of urea in a alkali soil with pH more than 10 is delayed as compared to normal soil.

The optimum N dose for medium duration, high yielding varieties of rice and wheat in alkali soils has been found to be 150 kg N per ha at Karnal, Kanpur and Indore. For short duration rice varieties, the optimum nitrogen dose is about 120 Kg N per ha. In order to reduce loss of N in volatilization and dentrification in alkali soils, N should be applied in rice and wheat crops in three splits, half or one third at transplanting / sowing and remaining in two equal splits at 3 and 6 weeks after transplanting / sowing. In case of sub-optimal dose of 80 Kg N per ha, it should be applied half at 3 weeks and remaining 6 weeks of crop age. In case of low dose of 40 Kg N per ha, the whole dose should be top dressed after 3 weeks of crop age.

Besides adding plant nutrients, application of organic matter brings out improvement in the chemical, physical and biological properties of the soil. Application of FYM and other organic manures hasten the reclamaition processes of alkali soils through solubilisation of native calcium as a result of chemical action of organic acids produced during decomposition of humus.

Among the several green manuring plants, *dhaincha* (*Sesabania aculeata*) during summer produced rice yield as much as under 80 Kg N per ha from fertilizer, thereby saving half of the dose of N in rice crop (Dargan *et al.*, 1975). The field trials conducted in India have revealed that among the several green manuring plants, *dhaincha* (*Sesbania aculeata*) has been found to be most beneficial in alkali soils. The best results were obtained when rice was transplanted immediately after the burial of *dhaincha* green manure crop.

Phosphorus

Under high soil pH conditions, availability of P is more and hence, no response to crops to P application has been found in the early phase of reclamation, but gradually soil gets improved and P availability decreases.

Alkali soils of Indo-Gangetic plains contain high amount of extractable P while in non-alkali calcareous soils water soluble P is rendered insoluble P in a few hours as it changes into insoluble calcium phosphate (Chhabra et al, 1980). In a number of field trials in India, no response of P application to rice in a rice-wheat rotation in a mild alkali soil is reported which is attributed to shallow root system of rice which fails to make use of higher P status of lower soil depths unlike wheat which is able to tap P from lower depths.

Potassium

No response of K has been reported in alkali soils under Indian conditions. A number of studies show reduced uptake of K by plants due to high Na and deficiency of available Ca in alkali soils. Consequently alkali soils do not respond to applied K fertilizers.

Calcium

Reduced availability of Ca is most limiting factor in crop production in alkali soil. Alkali soils are deficient both in soluble and exchangeable Ca at the result of which plants growing in alkali soils often succumb to calcium deficiency.

Micro-nutrients

In highly alkali soils, the availability of micro-nutrients, like Zn, Mn and Fe is very much depressed. Alkali soils contain as much total Zn as normal soils (40 to 100 ppm Zn) but still show severe Zn deficiency since Zn availability to plants is invariably related to soil reaction rather than to total Zn content of the soil. Field studies of Singh et al., (1987) showed that soil amended with 10 to 15 t of gypsum, 10 to 20 Kg of $ZnSO_4$ per ha was enough to meet Zn requirement of rice and wheat.

Since iron solubility is also conditioned by the soil pH, it is most commonly deficient micro-nutrient in alkali soils. Reduced soil condition brought about by sub-mergence and condition of organic materials help in mobilising the soil iron to meet plant needs. Excess water should be applied to rice nurseries when they show iron deficiency symptoms. Spray of FeSO may be done on upland crops showing bicarbonate induced iron deficiency.

Salts, like $NaHCO_3$ depresses the availability of Mn. Manganese may get reduced and leached from the upper layers under submerged conditions causing deficiency. Due to auto-oxidation, it is very difficult to correct Mn deficiency. However, in upland crops spray application of $MnSO_4$ may be helpful to make up deficiency.

(4) Water management

Water management for crop production in alkali soil is entirely different from normal soil because the water intake rate of alkali soils may be as low as 0.2 cm per day compared with 12.0 cm per day for normal soil (Singh et al., 1998). In alkali soils, water stagnation following rain storm or heavy irrigation, continues on the soil surface for a long period until it evaporates and pentrates hardly a few centimeters in the soil profile. During summer months, the soil surface gets dried up very quickly without any change in water content below 15 cm depth. This indicates a very poor upward water flux to replenish the water loss taking place on the surface.

Low permeability of alkali soils make them suitable for cultivation of rice, a semi-aquatic plant, which needs a submerged moisture regime for optimum yield. Experiment conducted at the Central Soil Salinity Research Institute, Karnal, Haryana revealed that application of 5 ± 2 cm of irrigation after a day of disappearance of ponded water, produced as much grain yield as a treatment with continuous submergence. This practice saved 20 per cent of irrigation water. Delaying the irrigation to four days after disappearance of ponded water though saved 60 per cent water but caused significant reduction in yield (Randhawa and Rajput, 1986).

The yield of wheat crop was significantly higher when first irrigation at crown-root initiation (CRI) stage was given at 30 days after sowing (DAS) but 21 to 24 DAS in alkali soils with 38-50 ESP (Sharma et al., 1985). Significantly higher relative growth, grain yield and water use efficiency was observed with increase in irrigation frequency from 0.6 IW/CPE (irrigation water / cumulative pan evaporation) ratio to 1.0.

Five irrigation in wheat scheduled at CRI, tillering, jointing, milking and dough stages resulted in higher yield, whereas under deficit water supply conditions for irrigations at CRI, jointing, flowering and milking stages besides pre-sowing irrigation would be sufficient to obtain reasonably high yields of wheat (Sharma et al., 1990). During *rabi* season, mustard crop responded up to 0.8 IW/CPE ratio in semi-reclaimed alkali soil.

REFERENCES

Abrol, I. P. and Bhumbla, D. R. 1971. Saline and alkali soils in India - their occurance and management. Paper presented at FAO/UNDP Seminar, Soil Salinity Research. FAO World Soil Resource Report No. 41: 42-51.

Abrol, I. P., Dargan, K. S. and Bhumbla, D. R. 1973. Reclaiming Alkali Soils, Bulletin No.2 Central Soil Salinity Research Institute, Karnal (Haryana), India, pp:58.

Abrol, I. P. and Fireman, M. 1977. Alkali and Saline Soils - Identification and Improvement for crop production. Bulletin No. 4. Central Soil Salinity Laboratory, Karnal (Haryana), India, pp: 8 - 32.

Abrol, I. P. Bhumbla, D. R. 1979. Crop response to different gypsum application in a highly sodic soil and tolerance of several crops to exchangeable sodium under field conditions. Soil Science, 127:79-85.

Abrol, I. P., Chhabra, R. and Gupta, R. K. 1980. A fresh look at the diagnostic criteria for sodic soils. **In**. Proceedings of International Symposium on Soil-affected soils, Central Soil Salinity Research Institute, Karnal (Haryana), India, pp: 142-147.

Agarwal, R. R. and Gupta, R. N. 1968. Saline-Alkali soils in India. ICAR Technical Bulletin (Agric Series) No. 15. ICAR, New Delhi, India, pp: 1 - 123.

Bernestein, L. 1964. Salt tolerance of plants. USDA Agriculture Information Bulletin No. 283. US Department of Agriculture, Washington D.C., USA, pp: 23.

Bhargava, G. P., Abrol, I.P. and Bhumbla, D. R. 1976. Journal of Indian Society of Soil Science, 24: 81-83.

Bhumbla, D. R. 1972. Reclamation and utilization of salt affected soils. Indian Farming, August 1972, pp: 151-154.

Bhumbla, D. R. 1977. Proceedings Indo-Hungarian Seminar on Management of Salt Affected Soils. Central Soil Salinity Research Institute, Karnal (Haryana), Indo. pp: 14-19.

Branson. 1960. Lectures at the Second Regional Training Courses on Soil Salinity, UNESCO, Lahore.

Chapman, V. J. 1976. Mangrove vegetation. J. Cramer, pp: 447.

Chhabra, R., Abrol, I.P., Dargan, K. S. and Gaul, B. L. 1980. Save on phosphatic soils reclamation. Indian Farming 30 (3): 13-15.

Chhabra, R. 1986. Soil Salinity and Irrigation Water Quality. Lauven: Centre for Irrigation Engineering, Catholic University of Leuven, and New Delhi: Oxford & IBH Publications.

Chhabra, R. and Kamra, S. K. 2000. Management of Salt affected Soils. International Conference on Managing Natural Resources for

Sustainable Agricultural Production in the 21st Century held on Feb. 14-18, 2000, New Delhi, India. Extended Summaries. Vol. 1. Invited papers, pp:47-49.

Chhabra, R. 2000. Amelioration of Alkali Soils for Crop Production. *In* Recent Advances in Agronomy (*eds.* G. Singh, J. S. Kolar and H. S. Sekhon). Indian Society of Agronomy, IARI, New Delhi, pp: 213-237.

Dargan, K. S. and Chillar, R. K. 1973. Technique of sugarbeet and vegetables in saline alkali soils. Indian Farming, 22 (12): 13-16.

Dargan, K. S., Gaul, B. K. and Abrol, I.P. 1973. Growing sugarcane in saline alkali soils. Indian Farming, 23 (2): 9-10.

Dargan, K.S., Chillar, R. K. and Bhardwaj, K. K. R. 1975. In alkali soils green manuring for more paddy. Indian Farming, 25 (3): 13-15.

Dukshina murti, C., Michael, A. M. and Shri Mohan. 1973. Water Resources of India and Their Utilization in Agriculture. Water Technology Centre, IARI, ICAR, New Delhi, pp: 357.

FAO-UNESCO (Food and Agriculture Organisation - United Nations, Educational Scientific and Cultural Organisation) 1973. Irrigation, Drainage and Salinity. An International Resource Book. Hutchinson, London, pp: 510.

Hayward, H. E. and Wadleigh, C. H. 1951. Plant Growth on Saline and Alkali Soils. Advances in Agronomy, 1: 1-38. Performance of tropical forage grasses with different methods of planting in alkali soils under rainfed conditions. Indian Journal of Agronomy, 37:527-532.

Kumar, A. and Abrol, I. P. 1986. Grasses in Alkali Soils. Bulletin No. 11. Central Soil Salinity Research Institute, Karnal (Haryana), India.

Mass, E.V. and Hoffman, G. J. 1977. Crop salt tolerance current assessment. Journal of Irrigation and Drainage, Division, ASCE 103: 115-134.

Massoud, F.I., 1974. Salinity and alkalinity as soil degradation hazards. Page 21 in Food and Agriculture Organisation, FAO/UNDP Expert Consultation on Soil Degradation, 10-14 June 1974. FAO, Rome.

Puntamkar, S.S., Mehta, P.C. and Seth, S. P. 1971. Note on inducement of salt resistance in two wheat varieties by pre-soaking with different salts of varying concentration. Indian Journal of Agricultural Science, 41: 717-718.

Randhwa, N. S. and Rajput, R. K. 1986. Irrigation Water Management Research - National Prospective. National Seminar on Water Management - The key to Developing Agriculture (*ed.* J. S. Kanwar). Indian National Science Academy, pp:389-391.

Rao, K.V.C.K. and Dhrvanarayan. 1978. Hydrology and Drainage Condition of Coastal Saline Soils (Unpublished).

Richard, L.A. (Editor)/USSLF 1954. Diagnosis and Improvement of Saline and Alkali Soils. United States Salinity Laboratory Staff. Agric Handbook 60, USDA, Washington D.C., USA, pp: 1-160. Publication: Oxford and IBH.

Schoonover, W. R. 1952. Examination of soils for Alkali. University of California, Extension Service, Barkeley, California.

Sharma, D. K., Singh, K. N. and Chillar, R. K. 1985. Effect of time of first irrigation on yield of wheat in alkali soils. Indian Journal of Agronomy, 30: 459-463.

Sharma, D.P., Singh, K. N. and Rao, K.V. G.K. 1990. Response of wheat, Indian mustard and barley to soil salinity in field. Indian Journal of Agricultural Science, 60 (9): 588-591.

Sharma, D. P. and Rao K.V.G.K.1996. Recycling of drainage effluent for crop production. *In* Proceedings of the Workshop on Waterlogging and soil salinity in Irrigated Agriculture, 12-15 March, 1996, CSSRI, Karnal (Haryana), India, pp:134-147.

Singh, K. N., Kumar, A. and Sharma, D. K. 1998. Management of Problem Soils. *In*. Fifty years of Agronomic Research in India (*eds*., R. L. Yadav, P. Singh, R. Prasad and I.P.S. Ahlawat). Indian Society of Agronomy, New Delhi, pp: 255-270.

Singh, N. T. and Chhabra, R. 1990. Management of Salt Affected Soils. *In*. Agronomic Research Towards Sustainable Agriculture (*eds*. K. N. Singh and R. P. Singh). Indian Society of Agronomy. IARI, New Delhi, India, pp: 131-152.

Singh, N.T. 1992. Salt affected soils in India. *In*. Land and Soil (eds. T. N. Khoshoo and B. L. Deekshatulu), pp: 65-100, New Delhi: Har-Anand Publication.

Srivastava, V.C., Jha, R. K. Sharma, N. N. and Verma, U. N. 1979. Response of certain *rabi* (winter) crops to limited water supply under saline conditions. Journal of Soil and Water Conservation in India, 29 (1-4): 9-15.

Swarup, A. and Singh, K. N. 1989. Effect of 12 year rice-wheat cropping sequence and fertilizer use on soil properties and crop yields in sodic soil. Field Crops Research, 21:277-287.

Swarup, A. and Singh, K. N. 1993. Amelioration effect of gypsum, FYM and nitrogen in a sodic soil. Annual Report, Central Soil Salinity Research Institute, Karnal (Haryana), India.

USSLS (United States Salinity Laboratory Staff) 1958. Salt Problems in Irrigated Soils, USDA Bulletin 190, August 1958.

Uppal, H. L. 1961. Reclaiming Alkali Waste Lands, ICAR Bulletin No. 85, ICAR, New Delhi, India, pp: 1-40.

Uppal, H. L., Agarwal, R. R. and Kibe, M. 1961. Reclamation of saline and Alkaline lands. Farm Bulletin 66:1-30. Farm Information Unit, Directorate of Extension, Ministry of Food and Agriculture, Government of India, New Delhi.

Yadav, J.S.P. 1996. Extent, nature, intensity and causes of land degradation in India. *In.* Soil Management in Relation to Land Degradation and Environment (*eds*. T. D. Biswas and G. Naryansamy). Indian Society of Soil Science, New Delhi, India, pp:1-26.

CHAPTER 4

Eroded Soils

Soil is most indispensable gift of nature to man. Soil along with sunlight, air and water nourishes all forms of life on this planet. Productive soil is the source of human sustenance and security. The danger that beset the productive capacity of the soil is erosion which removes the top layer of the soil containing all the principal plant nutrients and leave it unfit for sustainable production.

Soil is a complex mixture of animal, vegetation and mineral matter lying over the rocky core of the earth at different depths. The heat of the sun, the rain drops, the severe blasts of the wind, the cold and the winter season, the roots of the plants, the insect life, the human life and bacterial life are all acting upon the crust of the earth resulting in tremendous biological, chemical and physical changes. As a combined result, the hard rocky surface of the earth is crumbling down and is being shattered to pieces which get further powdered. These shattered and powdered parts get displaced by wind and rain and are carried away from higher and undulating places by the flowing water of streams, rivulets and rivers and by roaring winds and are deposited on lower and flatter areas. According to Rama Rao (1962), the main causes of erosion of soil and depletion of land resources can be enumerated as follows:

1) Heavy destruction of the natural protective cover like trees and grasses by indiscriminate cutting down of trees, forest fires, indiscriminate and over-grazing, and burning of grasslands.

2) Malpractices in the use of land such as up and down cultivation (on the slope), successive growing of crops that accelerate soil erosion, removal of organic matter and other valuable plant-food material by burning and injudicious cropping practices and faulty method of irrigation.

SOIL PROPERTIES AFFECTING ERODIBILITY

Soil characteristics that enhance or retard erosion by water or wind are discussed separately as follows:

Soil erosion by water: Soil erosion by running water is only possible when there is excess water over soaking for free runoff. Thus, while considering erosion by running water soil profile should be considered as a whole and not only the surface soil. The soil or on of the lower soil horizons should be considered while considering soil erosion by water.

i) **Permeability of the surface soil**: It is governed by the periodicity and character of rainfall, gradient and length of the slope, tillage, texture, structure, adsorbed cations, presence of crust, pan or other compact layers, organic matter and kind and amount of vegetal cover.

ii) **Permeability of subsoil or lower layers**: It is governed by their texture, structure, adsorbed cations, presence of pans or impervious / compact layers and high water table.

iii) **Permeability of the profile**: It is governed by soil depth over bed-rock.

Hence, the permeability either of surface or subsoil is most important factor influencing erosion of a land. Lowering of soil permeability either due to some inherent soil properties or by super-saturation of soil with water sets in runoff and accelerated water erosion.

Surface soil: The factors, affecting the permeability of the surface soils, is not completely independent of each other but are complementary to each other.

i) **Periodicity and characters of rainfall**: When the rain falls heavily in short time, the soil is not able to soak all of it and flows over the land causing erosion. Likewise, long persistent drizzle may saturate the soil beyond its capacity to soak, thus resulting in flows and erosion. Big rain drops falling on a tilled field may choke the pores by spluttering of the soil particles reducing permeability and causing runoff and erosion.

ii) **Gradient and length of slope**: Higher the gradient of slope, more is the velocity of slope, more is the velocity of running water and more its cutting and eroding power. Lands with long length of slope and lower gradient will accumulate enough water causing greater velocity and more cutting and eroding power.

iii) **Tillage**: On untilled, hard, barren surfaces, less water will be soaked in the soil resulting in more runoff and erosion than on a tilled land. A land tilled along the contour will offer no resistance to running water and the erosion will be greater than on the same land when tilled across the contour.

iv) **Texture**: The permeability in heavy soils is much low as compared to light soils which causes more erosion. Likewise unisized textural soils like clays, silts and sands are more edodible than multisized textural soils like loam, sandy loam and clay loam.

v) **Structure**: Soils of the same texture but of granular and crumby structure has better permeability than blocky structure and blocky structure has better permeability than prismatic and platy structure in order of merit. Soils with higher waterstable aggregates remain open and permeable for longer period than soils with low amount of waterstable aggregates which disperse and get puddled.

vi) **Adsorbed cations**: Calcium soils have more waterstable aggregates and can retain good permeability than acid soils. Sodium saturated soils become hard and dry. They disperse and puddle easily when wet.

vii) **Presence of crust, pan or compact layers**: Presence of crust, clay pans or compact layer of tillage pan or cattle tracks reduce the entry of water in the soil and enhance erosion.

viii) **Organic matter**: Organic matter helps in the formation of better or more stable aggregates and thus, helps in keeping the permeability high.

ix) **Kind and amount of cover**: A land covered with vegetation break the impact of raindrops and check the puddling and spluttering of soil particles. They help in binding the soil mechanically. Organic matter added by the dead and decaying roots, fallen leaves and twigs help to keep the soil in better structure. Trees with undergrowth is best in this respect, followed by grasses, trees and lastly cultivated crops.

Subsoil: The effect of texture, structure and adsorbed cations on soil permeability is same in the case of subsoil as already discussed in case of surface soil.

Presence of high water table checks the downward flow of water and thus, acts as a hard pan in restricting drainage and permeability.

In the rock occurs near the surface, it restricts drainage in the same way as a hard pan and thus, lowering the permeability.

Soil erosion by wind: Contrary to the soil conditions favouring soil erosion by water, soil erosion by wind occurs under dry conditions and high velocity of winds. Soil properties favouring erosion by wind are briefly discussed below:

i) **Tillage**: As the soils are loosened by tillage, the blowing wind removes finer particles in suspension by the process of saltation.

ii) **Structure**: Soils with single grained structure or with aggregates of lower stability are more prone to wind erosion. Aggregates of lower stability get easily broken and dispersed by jumping sand particles releasing the finer particles, like clays and silt, which are blown off by wind.

iii) **Organic matter content**: Soils with low organic matter have low amounts of waterstable aggregates and thus, facilitates wind erosion.

iv) **Barren surface**: If the soil is covered with vegetation, the wind cannot act upon the soil particles and hence, there is no chance of carrying off of the particles by wind.

v) **Continuous dry weather**: In wet and moist soils the finer particles like clay and silts are held together. Under such conditions, saltation by jumping sands is not possible and there is no danger of soil erosion by wind, as in the case of dry weather.

SOIL EROSION

Since the dawn of time, soil has been on the move. Rain and wind have scoured away almost constantly at the earth's surface and transported soil particles from place to place. In this way stream channels have been carved out, river deltas have been built up, and whole landscape have gradually been transformed. But under a protective cover of grasses, trees, or other thick-growing vegetation, the rate of soil removal has always been exceedingly slow- no faster, generally, than the normal rate of soil creation. The favourable soil balance prevailing under natural conditions was disturbed almost from the moment when man first started to till the earth for food by clearing away the wild growth and breaking the ground surface with crude implements. But farming probably went on for centuries before soil erosion became a recognizable human problem. It was only when population pressure forced the cultivation of steep slopes or unstable soils that people began to realize dimly that the earth can rapidly waste away under the impact of rain and wind (USDA, 1964). Soil erosion is now recognised as a serious threat to mankind.

Definition: Soil erosion is the process of destruction of the soil and the translocation of the products of destruction by water and wind. In other words, soil erosion is the denudation of land surface by detachment and transportation of soil by the action of such natural agencies as water and wind. The rich virgin soils, that nature may have taken thousand of years to build are being lost in only a few years time.

Basic process: Detachement and transport are basic processes occuring on source areas, while transport and deposition are basic processes occuring in sink areas. Soil particles are detached by either beating action of raindrop impacts or flowing water. Mutchler and Young (1995) reported that individual raindrops strike the soil surface at velocities up to 9 m per second, creating very intense hydro-dynamic forces at the point of impact. Overland flow detaches and/or entrains soil particles when its erosive hydro-dynamic forces exceeds the resistance of the soil to erosion (Foster and Meyer, 1972). While detachment by raindrops occurs over a broad areas, detachment by flow is often concentrated in small definable channels (rills). Detached particles are transported by raindrop splash and run-off. The hydrological processes of rainfall and run-off drive the erosion. Factors which affect amount and/or rate of rainfall or runoff also affect erosion. The deposition of eroded sediments takes place at the base of concave slopes, strips of vegetation, flood plains and reservoirs with reduction in flow velocities (Dhruva Narayana, 1993).

EXTENT OF PROBLEMS

Soil erosion is a major constraints in crop production in the highly diverse agro-ecological regions of India. High intensity rains during the monsoon season cause excessive runoff and soil loss resulting in impoverished and moisture- deficient soils. Soil erosion accounts for 87 per cent of total area affected by various forms of land degradation in India. The National Bureau of Soil Survey and Land use Planning, Nagpur has estimated earlier that the total soil erosion (both water and wind) has affected 166.4 million ha area, which accounts for 49.4 per cent of the total area of the country. The soil loss by erosion on unprotected land is often as much as 50 tonnes per acre and may go as high as 1200 tonnes fertile soil per acre.

Wind erosion is prevalent in arid and semi-arid regions of the country. It covers 28,600 square Km in Rajasthan, Haryana, Gujarat and Punjab, out of which 68 per cent is covered by sand dunes and sandy plains. It is estimated that 58 per cent area of Rajasthan, 51 per cent in Gujarat, 17.8 per cent in Haryana and an area of similar size in Punjab are affected by wind erosion.

Erosion by water is most serious process of land degradation and desertification in the country. On an average 5,334 million tonnes of top soil is lost annually (16.4 tonnes per ha per year) through water erosion of which about 1600 million tonnes of top soil representing 30 per cent of total eroded area is lost permanently into the sea, 10 per cent gets deposited in reservoirs reducing their capacity by 1 to 2 per cent every year, and remaining 61 per cent is displaced from one place to another. The annual water erosion rates vary from less than 5 tonnes per ha per year for dense forest, snow land cold deserts and arid regions of western Rajasthan to more than 80 tonnes per ha per year in Shivalik regions. Sheet erosion affects red soils comprising alfisols, ultisols, and oxisols (4 to 10 tonnes per ha per year) and black soils constituting vertisols and vertic soils (11 to 43 tonnes per ha per year). Gully erosion seriously affects 13 million ha with rates more than 80 tonnes per ha per year. Runoff erosion has resulted in extended and deep gullies in vast areas of Uttar Pradesh, Madhya Pradesh, Rajasthan, Jharkhand, Bihar, Gujarat, Maharastra, Andhra Pradesh, Tamil Nadu and Karnataka. Ravines which are net work of gullies associated with a river system affect 3.97 million ha in the country (Sharda and Jugal, 2006). On erosion-prone steep slopes of western Himalayas, about 50 per cent of the annual rainfall is lost as run-off, carrying with it 20 tonnes per ha soil and 18.7 Kg N, 17.0 Kg P_2O_5 and 41.4 Kg K_2O per ha (Narain *et al.*, 1994).

ESTIMATION OF SOIL LOSS

The universal soil-loss equation developed by Wischmeier and Smith (1965) and later refined (Wischmeier and Smith, 1978) predicts only the losses from sheet and rill erosion under specified condition. It enables to determine the land management erosion rate relationship for wide range of rainfall, soil slope, crop and management conditions and to select alternate cropping and management combinations that limit erosion rates to acceptable levels.

The universal soil loss equation (USLE) is:

$$A = RKLSCP$$

where,

A = the soil loss in tonnes per ha per year.

R = the rainfall erosivity factor which is the number of rainfall erosion index units for a particular location.

K = the soil erodibility factor which compares the actual erosion to that erosion from a clean, tilled, fallowed, uniform slope of 9 per cent and length of 22.13 m.

L = the slope length factor is the ratio of soil loss from the field slope length to that form a 22.13 m length under identical conditions.

S = the slope steepness factor is the ratio of soil loss from the field slope gradient to that from 9 per cent slope under otherwise identical conditions.

C = the cover and management factor is the ratio of soil loss from an area with specified cover and management to that from an identical area in tilled continuous fallow.

P = the support practice factor, is the ratio of soil loss with a support practice like contouring, strip cropping, or terracing to that with straight row farming up and down the slope.

Knowing the values of rainfall erosivity, soil erodibility and slope one can calculate the effectiveness of various erosion control measures to use, if one decides to introduce a cultivation system into an area with soil loss not exceeding an acceptable erosion rate.

TYPES OF EROSION

Normal or Geologic erosion: It is a natural process going on steadily but so slowly that ages are required for it to make any marked alteration in the major features of the earth's surface. It is both destructive and a constructive process. This process, under ideal constructive natural conditions, is both abrading (denuding) and aggrading (building). There is always an equilibrium between denudation and soil formation so that unless the equilibrium is disturbed by some outside agency the mature soil preserves more or less constant depth and character indefinitely.

Accelerated Soil erosion: The removal of the surface soil from areas denuded of natural protective cover, as a result of human and animal interference, takes place at a much faster rate than it is built up by the soil forming processes. It is the type of soil erosion that we are so seriously concerned. Nature requires normally 1,000 to 2000 years to built an inch (approximately 25.4 mm) of top soil but wrong farming methods and other misuse of soil may take only a few years to erode the same from lands of average slope, that means the loss of a thousand years of nature's work.

A. **Water erosion:** This can broadly be classified as (i) sheet erosion (ii) Rill erosion (iii) Gully erosion (iv) Land slides or slip erosion (v) stream bank erosion, and (vi) Sea-side or shore erosion.

i) **Sheet erosion**: It removes a thin covering of soil from large areas, often entire fields, more or less uniformly during every rain producing a runoff. Every times it rains, the running water becomes muddy and flows down to lower reaches, revulets or rivers and thus, million and millions of fine fertile soil particles which form the thin top layer of the soil are taken away by the flowing water. This type is very insidious in that, it keeps the cultivators almost ignorant of its ill-effects. As a result, it is neglected with consequent deterioration of land resulting in slowly yield decline.

ii) **Rill erosion**: In the next stage, when sheet erosion is allowed to continue unchecked, the silt laden runoff forms well-defined minute finger-shaped grooves ever the entire field. Such thin channelling is known as rill erosion. Year after year the rills slowly increase to get wider and deeper and thus, begins to hinder the movement of farm implements, reduces the actual area under crops, and results in declining crop yields.

iii) **Gully erosion**: This is third and advanced stage of erosion. When rill erosion is neglected, the tiny grooves develop into wider and deeper channels, which may grow into huge size. This is called gully erosion. Gullies are most spectacular symptoms of destruction through soil erosion which tend to deepen and wider with every heavy rainfall. They cut up large fields into small fragments and in course of time make them unfit for cultivation.

iv) **Land slides or Slip erosion**: This is caused by hydraulic pressure exerted or man-made slopes too steep for stability in wet condition. On becoming saturated with water, great masses of soil and loose rock lying on steep slope come down bodily. By this process, sometimes, an entire field or hill side may slide down.

v) **Stream bank erosion**: Stream- bank erosion usually manifests itself in two ways. The bank of the stream or river get eroded every year by the flowing water. In certain areas, the streams and rivers often change their course bit by bit every year. During sudden floods, the fast flowing river undercut the banks and erode the beds. Wherever there is a bend, the flowing water gets suddenly twisted and due to increased tension the outer bank caves in alongside. This eroded soil is deposited along the opposite bank nearby when the velocity is reduced. This process makes the stream meander. As a result quite often the stream or river changes their courses, divide lands into small fragments and cause formation of islands, screes, etc., in the stream course.

vi) **Sea or shore erosion**: The tidal waters of sea cause considerable soil erosion along the coast, particularly so during the rainy season, when the sea gets rough. The rivers in spate discharge their silt-laden muddy water into the sea. This usually results in silting up of the harbours and ports which consequently need dredging operation.

B. **Wind erosion**: Wind erosion takes place normally in dry areas devoid of vegetation where the wind velocity is high. The high velocity winds beat over the open lands with increasing force. The fine, loose and light soil particles blown from the surface of the land are taken miles and miles away. When the velocity of the dust bearing winds is retarded, coarser soil particles are deposited in dunes and the once fertile lands are rendered unfit for cultivation. In other places, fertile soil is drifted away by winds and the subsoil is exposed, as a result of which the productive capacity of the soil is considerably reduced.

MECHANISM OF EROSION

A. Water erosion: Soil erosion caused by rainfall is the result of the application of energy from two distinct sources, namely,

i) the falling rain drop, and
ii) the surface flow

The energy of falling rain drop is applied slantingly, vertically or horizontally along the surface of the ground. The chief role of the falling raindrop is to detach soil particles, whereas that of surface flow (outside the rills and gullies) is to transport soil. The falling raindrop also makes a major contribution to the movement of soil on unprotected sloping lands during periods of heavy impact by splashing large quantities down slope and by imparting transporting capacity to surface water by keeping it turbid. More than 100 tonnes of soil per acre are splashed by beating force of rain falling on the bare detachable soil.

Estimation of runoff: For estimating runoff three main methods commonly used are: (I) Rational method, (ii) Cook's method, and (iii) Hydrologic soil-cover complex method. The basic assumptions made in deriving the above methods are (a) rainfall occurs at uniform intensity for a duration at least equal to the time of concentration of the watershed and (b) rainfall occurs at a uniform intensity over the entire area of the watershed (Dhruva Narayana, 1996).

i) **Rational method**: This is the oldest, simplest and possibly the most consistent method expressed by an equation

$$Q = \frac{C/A}{360}$$

Where,
Q = design peak runoff rate in cubic meter per second ;
C = runoff coefficient ;
I = intensity of rain in mm per hour for the design recurrence interval and for duration equal to the time of concentration (t) of the watershed
A = Watershed area (ha)

The time of concentration (t) of the watershed is given by

$$t = 0.1947\,(K)^{0.77}$$
$$K = \sqrt{L^3}/H$$

where,
t = time of concentration (min)
L = maximum length of travel (m), and
H = difference in elevation between most remote point and outlet (m)

Values for the runoff coefficient (c) can be obtained from Table 13. The intensity (I) of rainfall for a recurrence interval (T) of 25 years is obtained from:

$$I = \frac{KT^a}{(t+b)^n}$$

TABLE 13: Values of C in rational Formula

Vegetative cover and slope	Soil texture		
	Sandy loam	Clay and silt-loam	Silt-clay
Cultivated land			
0 - 5%	0.3	0.5	0.6
5 - 10%	0.4	0.6	0.7
10 - 30%	0.52	0.72	0.82
Pasture land			
0 - 5%	0.10	0.30	0.40
5 - 10%	0.16	0.36	0.55
10 - 30%	0.22	0.42	0.60
Forest land			
0 - 5%	0.10	0.30	0.40
5 - 10%	0.25	0.35	0.50
10 - 30%	0.30	0.50	0.60

ii) **Cook's method**: This method is suitable for watershed up to about 400 ha. By this method, the runoff characteristics of a watershed are examined under four categories of relief, soil infiltration, vegetative cover and surface storage. Runoff curves are entered with the drainage area and the ΣW, and a value of peak runoff for 10-year-return period is obtained. The peak runoff value is modified by multiplying it with the rainfall factor R obtained from frequency factor for various zones F and shape factor S.

iii) **Hydrologic soil cover complex number method**: This method developed by Ogrosky and Mockus (1957) determines peak rate of runoff for small watershed by synthesizing information about flow characteristics, physiographic factors and soil cover data. Ogrosky and Mockus (1957) suggested the use of 6-hour rainfall for the design frequency for small watersheds. It involve conversion of rainfall to runoff depth and the peak rate of runoff is obtained by the following relationship:

$$q\ peak = \frac{0.0208 \times A \times Qd}{Tp}$$

Where
q peak = peak rate of runoff, cumec;
A = area (ha) ;
Qd = runoff depth, cm ;
Tc = time of concentration (hrs)
Tp = time of peak (hrs)
$= 0.6\ Tc + \sqrt{Tc}$

Graphical solution of the above equation has been developed. This method can be used for estimation of water yield as well as peak rate of runoff from areas above 50 ha.

Estimation of direct runoff from rainfall: The rainfall - runoff relationship is given as:

$$Q = \frac{(P - Ia)^2}{(P - Ia) + s}$$

Where,
Q = actual runoff
P = the rainfall
S = potential maximum retention
Ia = Initial abstruction during the period between the beginning of rainfall and runoff in equivalent depth over the catchment condition of region.

To shows the rainfall-runoff relationship graphically, S values are transferred into curve number (CN) by the following equation:

$$CN = \frac{25400}{245 + s}$$

Using the above, equations developed are:

$$Q = \frac{(P - 0.3\,S)^2}{(P + 0.7S)} \quad \text{.................................... (1)}$$

$$Q = \frac{(P - 0.1S)^2}{(P + 0.1S)} \quad \text{.................................... (2)}$$

The equation (1) is applicable to all soil regions of India except the black soil areas referred to Hydrological Soil Groups.

Graphical solution of the above equation has been developed. The Hydrologic Soil Cover Complex Number Method can be used for estimation of water yield as well as peak rate of runoff from areas above 50 ha.

B. Wind erosion: The wind is responsible for three types of soil movement in the process of wind erosion. They are

i) **Saltation**: The major portion of the soil carried out by wind is moved in a series of short bounces called saltation. The soil carried in saltation consists of fine particles ranging from 0.1 to 0.15 mm in diameter. Saltation is caused by the direct pressure of the wind on soil particles. The initial rise of grain in saltation is generally in a vertical direction.

ii) **Suspension**: Very fine particles less than 0.1 mm in diameter are carried into suspension, being kicked up into the air by the action of particles in saltation. Suspended material by the movement of wind is carried long distances from the original location and is, thus, a complete loss to the eroded area.

iii) **Surface creep**: Grains larger than about 0.5 mm in diameter but smaller than 1.0 mm are too heavy to be moved in saltation but are pushed or spread along the surface by the impact of particles in saltation to form surface creep.

About 90 per cent of the total soil movement in wind erosion is below the height of 30 cm, and about 50 per cent of it is within 50 mm of the ground level. The control of wind erosion is mainly based on reduction or elimination of the movement in saltation.

FACTORS OF EROSION

Soil erosion by water

Soil erosion by water action is influenced greatly by (i) rainfall erosivity, (ii) percentage slope, (iii) length of slope, (iv) soil erodibility, and (v) nature of ground cover.

i) **Rainfall erosivity**: Erosion is directly a function of rainfall and runoff. The force required for particle detachment is provided by rainfall and is a function of intensity, drop size and wind accompanying a rainstorms. The kinetic energy of rain has been recognised as a major causative factor. The high intensity of tropical rain is associated with large drop size causing splash erosion due to beating action of rains on soil surface. Rainfall intensity, duration and frequency influence the rate and volume of runoff. A light rain which can be easily absorbed in the soil causes no runoff and soil loss. As intensity of rain increases, more rain falls than can enter the soil. Rainfall of long duration and greater frequency increases both the total runoff and soil loss.

ii) **Percentage slope and length of slope**: The speed and extent of runoff depends on the slope of the land. The greater the slope, the greater is the velocity of flow of the runoff water. According to the law of falling bodies, velocity varies as the square root of vertical drop. Hence, if the land of the slope is increased four times, the velocity of water flowing on the slope is approximately doubled. If the velocity of runoff water is doubled, its energy, i.e. erosive power is increased four times as the later varies as the square of the velocity. Similarly, the quantity of material of a given size that can be carried varies as the fifth power of velocity, and the size of the particles that can be transported by pushing and rolling varies as sixth power of the velocity. There is, thus, a rapidly increasing rate of soil loss as the slope of the field becomes steeper, if the percentage is not greatly affected by increase in slope steeper than 3 per cent depending upon the soil type.

iii) **Soil erodibility**: The type of soil, i.e., its permeability, texture, structure and organic matter content affect soil erodibility by water. Heavy or alkaline soils are highly susceptible to erosion. They get easily puddled and very quickly form an impermeable soil layer near the surface and consequently retards infiltration. Soil left in loose and pulverized condition is particularly liable to erosion through sheet wash and gullying.

iv) **Nature of plant cover**: When rain falls on a surface covered by a thick mantle of plants, most of it quickly percolates into the soil. Areas not protected with thick cover of plants are unable to absorb water effectively because the dashing rains shatter the soil surface, the fine soil particles go into solution or suspension and the thick mixture of water and soil quickly fills and closes the interstices in the soil, reducing infiltration and consequently increasing runoff and soil loss.

Soil erosion by wind

Wind also is an important factor that influences soil erosion. The effect of the velocity of wind decreases or increases according to the presence or otherwise of wind barriers in the form of grasses, shrubs and trees. The soil lost by wind erosion is possibly the richest soil as the humus is lighter than the soil particles.

A general functional relationship between different parameters that affect wind erosion has been developed into the wind erosion equation (Chepil and Woodruff, 1963; Woodruff and Siddoway, 1965).

$$E = f(I, K. C. L. V)$$

Where,

- I is soil erodibility index
- K is soil-ridge roughness factor
- C is climatic factor
- L is field length along the prevailing wind erosion direction, and
- V is equivalent quantity of vegetative cover.

Biological factors of erosion

There is hardly any accelerated soil erosion where the lands are left to nature. Soil erosion becomes a problem when man enters the picture. The indiscriminate felling of trees for the needs of human beings accelerate the process of soil erosion to a very great extent as the forests are denuded of the vegetative cover which holds the soil. Forest fire also cause similar results. Overgrazing of pasture by cattle and sheep denudes the land of its protective cover. In addition, shifting cultivation and other faulty cultivation practices accelerates the erosion process. Footpath used by human and cattle develop into rills and then into gullies.

MANAGEMENT OF ERODED SOILS

The basic factor in the application of erosion control is the proper utilization of the land according to its capabilities. The concept of land-capability classification has been developed in the USA for soil conservation of farm lands. This concept has been adopted in India also by the All India Soil and Land Use Survey Organisation for similar purposes.

Land capability classification

Land capability classification is a systematic arrangement of different kinds of land according to those properties that determine the ability of land to produce crops on a virtually permanent basis.

In order to classify the land according to their capabilities, the chief soil characteristics which are taken into account are texture, depth, permeability, salinity and alkalinity of soil and subsoil. The important associated land features are slope of land, effect of past erosion, natural soil drainage, frequency of overflow, etc.

The eight land capability classes range from the bast and most easily farm land (class I) to land which has no value for cultivation, grazing or forestry, but may be suited for wildlife, recreation or for watershed protection (class VIII). Classes I, II and III include land suited for regular cultivation (USDA, 1964).

Class I: This includes land suited for regular cultivation. It is a very good nearly level land, with deep, easily worked soils that can be cultivated safely with ordinary good farming methods.

Class II: This land is good but it has some limitations. It needs moderately intensive treatment if it is to be cultivated safely. Such treatment, for example, may be contouring and cover cropping to control erosion, or simple water-management operations to conserve rainfall.

Class III: This land is moderately good and can be used for cultivated crops regularly in a good rotation, if ploughed on the contour on sloping fields. It has limitations of such degree that intensive treatment is necessary which may be terracing or strip cropping to control erosion, or intensive water management on flat, wet areas.

Class IV: It is fairly good land, but its safe cropping use is very limited by natural features such as slope, erosion, adverse soil characteristics, or adverse climate. As a rule, its best use is for pasture or hay, but some of it may be cultivated occasionally with proper safeguards.

Class V, VI and VII are not suited for any cultivation but may be used for grazing or forestry according to capabilities. **Class VIII** land is suited for wildlife, recreation, or watershed protection. This land is characterised by such features as extreme steepness, roughness, stoniness, wetness, sandiness, or erodibility.

Soil conservation measures

Soil conservation is the use of land, within the limits of economic practicability, according to its capabilities (the way nature made the land) and its needs (its condition resulting from the way man has used it) in order to keep it permanently productive. All adoptable measures required to achieve permanent productivity of the soil constitute the tools for conservation, whether used singly or in various adoptable combination (Bennett, 1955).

The practical methods of soil conservation broadly falls into two important classes: A. Biological and B. Mechanical. The biological measures may again be divided into (a) Agronomic (b) Agrostological (c) Afforestation. Each of these conservation techniques should supplement and strengthen the others in an integrated programme. The basis of all conservation measures proposed to be adopted by man should be patterned as closely as possible after Nature's own way of conservation. Greatest emphasis should, therefore, be laid on the biological methods of erosion control.

A. Biological

(a) Agronomic practices

Although agronomic practices are considered as a second line of defence in soil and water conservation, these are of immense importance. Research results obtained from a range of soils and climatic environments have indicated that the best method of erosion control is good soil and crop management. Preference is given to soil and crop management techniques that minimize water runoff through improvement in soil structure and water infiltration. The management practices which contribute to the conservation and productivity of agricultural lands go to form what is known as conservation farming.

Contour farming: A contour line connects points of the same level and it runs across the slope of the land. Contour farming consists in doing all tillage operations including the sowing of seeds and planting in rows along the contour lines instead of the straight lines or in rows running up and down the slope. That is, the rows run more or less parallel to each other but across or right angles to the direction of the slope. The plough furrows are all level in this case and holds the light rains till they are absorbed by the soil. Even in the event of heavy rains, the numerous stems of a dense stand of plants impede the flow of moving water, thus reduces runoff, saves more moisture for crop production, reduces soil losses and increases crop yield. All terrace lands should be farmed on the contour. Contour farming combined with strip cropping maintain a more uniform distribution of water. For slopes up to 1 to 2 per cent, contour farming alone is effective. On steeper slopes and gullied lands, it ought to be supported by terracing, strip cropping and other measures to be effective.

i) **Tillage:** Fine pulverizing and deep and excessive tillage destroys soil structure in the long run and often lead to excessive soil erosion. Moreover, a thin impermeable soil layer formed just beneath the plough layer decreases the infiltration capacity, which in turn increases water runoff. Crust formation is also an important factor in decrease of infiltration rate of a plough bare soil surface. Shallow ploughing has been found adequate to kill weeds and enable soil to absorb rainfall. Contour tillage is essential ingradient of conservation farming. It helps in creating obstructions to the flow of water at every furrow results in a more uniform distribution of water, more infiltration of water and less runoff and erosion. For the no-tillage system to be effective for crop production in eroded soil, considerable information is needed for different soils, crops, and ecological regions. Special requirement of the no-tillage system are economical methods of weed control, suitable planting equipment, balanced fertilizer application, suitable crop combinations and cropping sequences, and an adequate amount of crop residue for mulch.

ii) **Mulching:** Crop residue mulch is one of the most effective measure against soil erosion. Mulch prevents the direct impact of raindrops on soil aggregates, maintains pore space continuity and high infiltration rate, and enhances crop growth to provide an early ground cover through improvements in soil temperature and moisture regimes and other physico-chemical properties. Tillage that leads to the surface soil cloddy and mulched with crop residue (stubble) is an effective measure to minimise soil erosion and to conserve moisture. Interculturing operations kill weeds and produce few centemeters of thick soil mulch which helps to reduce moisture evaporation from top soils.

iii) **Conservation tillage**: Conservation tillage is any tillage and planting system that maintains at least 30 per cent of the soil surface covered by residues after planting to reduce water erosion, or where wind erosion is primary concern maintains at least 1000 kg per ha of flat small grain residue equivalent on the surface during the critical wind erosion period (CTIC, 1995). Some relevant type of conservation tillage systems include (Lal and Kimble, 2000):

a) **No tillage**: In this system, the soil is completely undisturbed from harvest to planting except for planting seeds in soil, plant and nutrient application. Weed control is achieved primarily through herbicides.

b) **Reduced tillage**: The soil is disturbed as little as necessary prior to sowing by implements, such as chisels, disks or any other appropriate equipment (e.g., cultivators, sweeps, blades) to brake the crust, loosen the compact layer and obtain good crop stand. Weed control is achieved by herbicides and secondary tillage operations.

c) **Mulch tillage**: The soil is disturbed prior to sowing only by those equipment which do not bury the crop residue in order to provide an adequate amount of crop residue mulch on the soil surface. Desired weed control is accomplished with herbicides and/or cultivator.

d) **Conventional tillage**: It involves the plough based tillage method of soil inversion, and those which leave less than 15 per cent residue cover after planting or less than 500 kg per ha of small grain residue equivalent throughout the critical wind erosion period. This system involves the combination of primary, secondary and tertiary tillage operations.

e) **Traditional tillage**: It is indigenous tillage method based on specific tools developed and used in the region. Tillage may be performed by manually operated tools, animal drawn equivalent and/or motorized implements. The basic problems with most traditional tillage systems in India are:

   i) all crop residues including the roots and stumps are removed and/or grazed,

   ii) most soils are crusted, compacted, and have massive structures that cannot be improved by shallow traditional tillage systems.

   iii) low water infiltration rate leads to surface ponding and high losses of water due to surface runoff and evaporation, and

   iv) the low level of soil organic carbon pool is further depleted by mineralization and accelerated soil erosion.

iv) **Water harvesting and runoff recycling:** In regions of water scarcity, water supplies for crop production can be developed by capturing and storing a portion of the rainfall that is otherwise lost. Increased yields from watersheds in low rainfall areas due to harvested water which is substantially potential for irrigation and livestock use. The portion of rainfall which moves down to stream, channel, river or ocean as surface or sub-surface flow is termed 'runoff'. The runoff can be harvested from field, stored in a pond or tank and recycle it for life saving or supplemental irrigation to crops.

The technology of water harvesting consists of collecting and storing water from land treated to increase runoff. Runoff from the catchment can be increased by building contour ditches to collect hillside runoff, by clearing vegetation and smoothening the land surface, by treating soil surface to reduce the infiltration rate and by covering the soil surface with impermeable sheeting. Runoff from bare and unproductive soil can often be increased by dispersing the aggregates particles with sodium salts and hydrophobic chemicals to reduce infiltration. Impermeable catchment may be constructed by spraying bitumen compounds on the soil surface. The soil surface is also treated by spraying with a homogenized mixture of asphalt and water.

In order to reduce seepage in reservoirs, plastics, artificial rubber and chemicals could be used for lining and covering storage reservoirs to stop seepage and evaporation losses. Waterborne asphalt emulsions are effective in preventing seepage from reservoirs. Mixtures of bentonite and sodium carbonate or sodium chloride provide effective low cost sealants for form ponds.

Unrestricted evaporation may result in loss of 25 per cent or more of the water stored in tanks. The surface of the reservoir may be covered with a fully compressed mono-molecular layer of fatty alcohol series of compounds or solid floats to retard evaporation.

Rain water harvesting consists of in-situ and ex-situ harvesting. In-situ rain harvesting can be achieved by increasing infiltration rate with the help of deep ploughing, profile modification, vertical mulching and by keeping soil surface rough. In arid and semi-arid areas, the low and erratic rainfall normally occurs with high intensity of short duration causing high runoff and poor soil moisture storage resulting in loss of 50 to 60 per cent of rain water in India. The surface runoff from an area can be increased by reducing the infiltration capacity of the soil through vegetation management, cleaning, sloping surface vegetation and reducing soil permeability as discussed above. The ex-situ water harvesting technology include roof-top collection, dug out ponds/ storage tanks, gully control structures/ check dams, water harvesting dams, precolating tanks/ ponds, sub-surface dams/barriers, etc. Roof-top harvesting is recommended only in areas of high rainfall intensity spreading over the large parts of year, e.g. Himalayan areas, north Andaman and Nicobar islands.

v) **Choice of crops and cropping sequence:** Soil erosion depends not on what crop is grown but on how it is grown. An ill managed corn crop produced 15 times more soil loss than a well managed corn under identical conditions. Even a well-managed corn produced less soil loss than an ill managed pasture (Hudson, 1957). Some crops are prone to erosion because the way they are grown, like clean-tilled row crops. Other close growing crops like grasses and good canopy producing crops like legumes suffer less erosion. Grasses have been found to control soil erosion, improve soil structure and all this is due to good canopy and excellent binding force shown by the grass roots. Next to grass may be legume roots which help in improvement of soil structure due to higher root mass, and root secretions, helping in binding soil particles. Thus, legumes are considered erosion resistant crop, while cereals are considered erosion permitting crops because of manner in which they are cultivated.

vi) **Cover crops:** The continuous loss of soil from the cultivated lands means reduced fertility and consequent increased expense, if crop production is to be maintained. Soil is more susceptible to erosion when fields are bare of vegetation particularly on slope and soil is low in organic matter content.

Any crop while serving as a solid ground cover, whether or not specifically planted for that purpose, is a cover crop. Thus, a growing grain crop, grasses in pastures and crops planted for turning under as green manure are cover crops. In other words crops planted especially for the purpose of checking soil erosion, adding organic matter to the soil, and improving soil productivity are generally termed as cover crops.

The principal advantage of cover crops are: (i) reduced runoff of rain and thus, conserves rainfall, (ii) prevents excessive erosion of the land, (iii) improves soil tilth by adding organic matter to the soil and by loosening up the surface of the soil with deep-growing roots, (iv) diminishes leaching of available plant nutrients, especially of nitrates, (v) protects terraces and other mechanical erosion control devices, (vi) increases the water absorbing capacity and the infiltration of water into the soil, and (vii) increases yield of cultivated crops and provides pastures all the year round.

Crop cover or canopy is most important factor affecting soil loss dramatically. Generally legumes provide better cover and protection to land. Consequently, the soil and water loss is reduced by growing legumes. Studies conducted by Tejwani *et. al.,* (1975) revealed that cowpea provided an excellent cover to ground followed by other legumes like velvet beans, sumhemp, *dhaincha* (sesbania), greengram and blackgram.

vii) **Cropping sequences:** The continuous growing of the same crop on the same land, irrespective of the nature of crop grown, leads to considerable breakdown in the structure of the soil, depletion of plant nutrients from the soil, enhancing sheet and gully erosion due to continuous clean cultivation of row crops and reduction in crop yields. Cropping sequences that provide early and continuous ground cover permit less erosion. A suitable cover crop used in a crop rotation helps to improve soil physical and chemical properties. Infiltration rate of structurally degraded soil can be rapidly improved with 1 to 2 years of a deep rooted cover crop, such as pigenpea (*Cajanas cajan*), stylo (*Stylosanthes guianensis*), valvet beans (*Psophocarpus palustris*), etc. (Lal *et.al.,* 1978, 1979). Through improvements in soil physical properties and prevention of raindrop impact, cover crops in cropping sequences prevent runoff and erosion during their growth cycle and for food crops grown on the same land in sequence.

Cultural practices (e.g. balanced fertilizer, insect and pest control, optimum plant population, early planting, etc.) that favour good crop growth also minimise losses due to runoff and erosion (Vasudeiaiah *et.al.,* 1965).

Mixed and relay cropping, widely practiced in tropics provide a continuous ground cover. A crop combination with multistory canopy structure should be effective against erosion control. Advantages of mixed cropping are good crop cover, feeding of crop from different soil layers and under rainfed conditions an assurance to farmer against a total crop failure.

viii) **Strip cropping:** This consists of growing different crops in relatively narrow strips across the slope of the land in such a way that rows of erosion permitting (clean tilled and row crops) crops are put alternatively with the rows of erosion resisting crops (closed grown crops). This practice reduces the velocity of runoff water and the silt gets deposited within the erosion resisting crop strips. The dense foliage of these strips prevents the detachment of soil by splash caused by raindrops.

a) **Contour strip cropping**: It is the growing of erosion permitting crop in strips of suitable width across slopes on the level or contour.

b) **Field strip cropping**: It is the planting of farm crops in a more or less parallel strips across fairly uniform slopes but not on exact contour.

c) **Wind strip cropping**: It consists of planting tall growing crops, such as sorghum or maize in straight and long but relatively narrow parallel strips laid right across the direction of prevailing wind regardless of contour in between low growing crops.

d) **Permanent or annual strip cropping**: In this case, strips are established on steep or highly eroded slopes and these do not form a part of rotation as the case in the system of strip cropping. These strips may be of permanent legumes or grasses.

e) **Buffer strip cropping**: Buffer strips are normally the grasses and other erosion resisting vegetation and are permanent strip of varying width for the effective control of soil erosion and restoration of soil fertility.

ix) **Seeding / planting of crops**: It should be so adjusted that maximum canopy synchronizes with high intensity rains, thus reducing impact of rainfall on soil. Contour cultivation and contour plantings are basic agronomic practices which should be adopted on slopy lands.

The sowing date should be so adjusted that by the time intensive rainfall takes place, there is enough of ground cover, which is possible by advancing date of sowing or even dry sowing in rainfed areas in anticipation of rains.

Optimum plant population and crop geometry are equally important to have an optimum crop cover. Dense plant population may be better but excess population may lead to severe moisture and nutrient stress, particularly under rainfed.

x) **Nutrient management**: Organic manures and green manures not only supply plant food like fertilizers, but help in improving physical conditions of soil, i.e. improving soil structure, infiltration rate and moisture holding capacity of soil. The fertilizer improves vegetative growth and canopy of crop and reduces soil erosion. The crop residues left in the field after harvest can have tremendous positive influence on soil and water conservation. Crop residue retention is a valuable asset for the reduction of soil erosion.

(b) Agrostological measures

Grasses are considered to be one of the best tools in soil conservation. The improvement of soil structure and maintenance of organic matter by grasses make soil resistant to detachment and transportation by wind and water. Grasses acts in the same way as legumes in controlling the soil and water loss. Strip of grasses laid in between growing field crops are highly beneficial in conserving soil and soil moisture, and improving the structure and productive capacity of the soil.

In places where erosion control by agricultural practices alone or in combination with mechanical means is not possible and where the land is very badly eroded, the best way to stop further destruction of the soil lies in the area being put under grasses.

Various grasses for different purposes found suitable (Dhruva Narayana, 1993) are as follows:

i) **Soil conservation only**: *Eragrostis curvula* (weeping love grass), *Tripsacum laxum* (Gautemala grass).
ii) **Soil Conservation and fodder**: *Festuca elatior* (Tall grass), *Phalaris tuberosa* (Hardinge grass), *Dactylis glomerata* (Cock's foot), *Lolium perenne* (Rye grass), Pusa Giant Napier (NB-21), *Agropyron semicostatum* (Wheat grass).
iii) **Grassed waterway**: *Paspalum dialatatum* (Dallis grass), *Cynodon suwanne* (Doob grass), *Agrostis tenuis.*
iv) **Bench terrace embankment**: *Eragrostis curvula*, *Dactylis glomerata, Digitaria brownii, Digitaria pentzi, Paspalum dilatatum.*
v) **Puetorican bench terrace formation**: *Tripsacum laxum*, Pusa Hybrid Napier (NB-17 and NB-21).

Grasses in stablisation of bunds and gully control: Grasses having good capacity for binding the soil together and also drought resistant are: Blue panic (*Panicum antidotale*), Napier grass (*Penninsetum purpurium*), Rhodes grass (*Chloris gayanna*), Marvel (*Dicanthium nodosum*), Pavane (*Sehima nervosum*) and Anjan (Cenchrus ciliaris). These grasses have been identified by the research conducted at Sholapur (Maharashtra). Similar observations made in the Nilgiri hills have indicated that *Phalaris tuberosa*, *Eragrostis amabilis*, Eragrostis curvula (weeping love grass), Cynodon dactylon (hariali), and *Dactylis glomerata* (Cock's foot) are very effective in binding the soil, stablising the risers of the bench terraces and sodding water channels. In Madhya Pradesh, *Thelepogan elegans*, *Pennisetum pedicellatum* amongst the annual grasses and *Chloris bournei*, Thin Napier (*Penninsetum polystachyan*) among the perennial grasses have been found suitable for similar purposes.

Ley farming: It consists of growing grasses in rotation with agricultural crops. Ley farming also helps in building up the structure of the soil and preventing soil erosion. This method can be adopted in places which abound in good pasture land.

Retiring lands to grasses: Pasture afford one of the most effective and economic means of holding and enriching the soil provided they are properly developed and managed. The outstanding requirements of good management in humid regions with adequate rainfall for abundant plant growth are (i) a sufficient supply of mineral plant nutrients, such as lime, phosphorus and potash for desirable plants to make enough growth to cover and protect the soil and to provide forage for livestock, and (ii) regulation of number and kind of livestock and the period of grazing (USDA, 1964). The vast patches of land in India are severely eroded. These areas should be put under pasture, properly manured with nitrogenous and phosphatic fertilizers. In order to avoid undue damage, rotational grazing should be adopted. Properly managed grass stands and pastures provide natural vegetative cover and thus, help in preventing soil and water loss, improving soil structure and productive capacity, stablising bunds, waterways and risers of terraces preventing floods and subsequent damage of low lying areas, and stablising shifting sand.

Regulated grazing: Rotational grazing should be adopted to avoid undue damage to the pasture as well as the land. Grazing should be prohibited where the slope is very steep and it should be limited in gentle slope. The tuft left behind after grazing are not able to arrest the flow of rain-water.

(c) Afforestation

Afforestation is most efficient agent for controlling soil erosion. Forest cover protects the land by intercepting rain and offering resistance to moving water and wind shielding the soil from erosion. As much as 25 to 30 per cent of total annual precipitation may be intercepted by a dense stand of forest cover. Forest litter serves as a protective layer by absorbing the impact of the falling rain drops and thus, protecting the soil particles to be eroded by splash erosion. Tree roots bind the soil and improves its porosity. The water is able to seep underground along the channels formed by the roots of the plant and by various animals and insects. The effect exerted by tree crowns in reducing wind velocity and thereby, controlling wind erosion is equally important. Similar reduction of wind velocity has been observed horizontally on the leeward side of shelter-belts and wind-breaks.

Floods are direct result of deforestation in catchment areas. Afforestation of steep slopes is the only feasible method of protecting cultivable fields in valleys by preventing flood and debris entering them.

Some of the species suitable for afforestation are *Tectona grandis*, *Shorea robusta*, *Dalbergia sisso*, *Acacia arabica*, *Atocarpas hirsata*, *Artocarpas integrifolia*, *Mangifera indica*, *Terminalia tomentosa*, *Terminalia arjuna* and *Dalbergia latifolia*.

The deep and narrow gullies are best controlled by putting them to permanent vegetation, Deep gullies are utilized as woodlands by planting suitable tree species. Gully heads and sides are stabilized by planting grass species. Reducing the slopes of gullies and directing the surface flow are essential pre-requisites for successful afforestation programme. Suitable species for afforestation of gully land are given below:

i) **Trees and shrubs**: *Acacia nilotica*, *Acacia catechu*, *Azadirachta indica*, *Albizia lebbeck*, *Dalbergia Sissoo*, *Prosopis chilensis*, *P. juliflora*, *Eucalyptus* species.

ii) **Grasses and legumes**: *Cenchrus ciliaris, Pennisetum pedicellatum, Saccharum spontaneum, Dicanthum annulatum, Cynodon dactylon, Hetropogan contortus, Stylosanthes* species.

(d) **Grass Waterways**: Grass waterways are essentially required on agricultural land, where a suitable natural drainage way is not available to drain off excess run-off safely at a non-erosive velocity. The cross-section of grassed waterway may be trapezoidal, triangular or parabolic. The suitability of grass species should be based on the density of the cover, ease of establishment and forage yield obtained.

B. Mechanical Measures

Mechanical measures of soil conservation are adopted to supplement the biological methods when the later are not adequate effective. The principles involved in the design of all mechanical structures to control erosion are: (a) increasing the time of concentration and thereby, allowing more runoff water to be absorbed and held by soil, (b) dividing a long slope into several short ones, so as to maintain a critical velocity for the runoff water, and (c) protection against damage due to runoff.

Some important mechanical measures are discussed below:

i) **Basin listing:** With the help of a special implement called basin lister, small interrupted basins along contour are formed to help in retaining rain water. It is specially effective on retentive soils having slopes milder than about 2 per cent.

ii) **Pan breaking:** In areas having hard pan sub-soil, the pan breaker is used at interval of approximately 90 to 180 cm on contour to improve percolation and drainage of rain water, thus reducing runoff and soil erosion.

iii) **Sub-soiling:** This method consists of breaking with a sub-soiler the hard and impermeable sub-soil to conserve more rain-water by improving the physical condition of the soil. This operation does not involve soil inversion, promotes more moisture penetration into the soil where the plant roots can penetrate with great ease. The sub-soiler is worked through the soil at a depth of 30 to 60 cm at a spacing of 90-180 cm.

iv) **Contour terracing:** The principle of controlling erosion by systematically intercepting surface runoff on sloping land has led to the use of farm terracing. When terraces are properly used and constructed, are adequately applicable to cultivated land. The ploughing parallel to terrace, so that a dead furrow falls in channel and a back-furrow on the ridge, keeps the channel open and builds up the ridge. Contour consists of constructing a series of properly spaced ridges or drainage channels. They are called 'level' terraces when constructed along contour and 'graded' if they are given a slight longitudinal gradient to promote safe discharge of excess water. The terraces are of four types.

a) **Channel terrace or graded bunding**: The broad channel terrance acts primarily to conduct excess rainfall from the fields at non-erosive velocities. It consists of a wise and relatively shallow channel excavated at suitable intervals on a falling contour with a suitable longitudinal gradient. The excavated soil is deposited as a wide, low ridge along the lower edge of the channel. This terrace is suitable for relatively impervious soil in heavy rainfall tracts. A high ridge is not desirable since it seriously interferes with tillage operations and increase construction. Graded bunds or channel terraces are recommended on mildly sloping lands (<6%) and high rainfall areas (>800 mm)

b) **Broad based ridge terrace**: It consists of a wide based, low embankment formed on contour by excavating soil from both side of it. It is more suited to low rainfall areas and soils that will absorb the accumulated rainfall fast enough to prevent damage to growing crops.

c) **Narrow based ridge terrace**: It consists of narrow based embankment constructed at intervals across the slope of the land either on a level that is along contour, or slightly graded longitudinally. The vertical spacing between bunds may ordinarily vary from 7 to 15 cm approximately. These bunds are constructed on agricultural lands up to a slope of about 10 per cent. As regards spacing between bunds it should not exceed 150 cm vertical drop or 67.5 m horizontal spacing whichever is less. Where the rainfall is more than 60 cm and moisture conservation is important, bunds may be constructed at shorter intervals.

d) **Bench terracing**: Bench terracing is one of the oldest mechanical methods of erosion control, having been used for many centuries in thickly populated countries where economic conditions necessiated the cultivation of steep slopes. The use of bench terraces on steep slopes not only retards erosion but it also makes cropping operations on these slopes possible and safe. This type of terrace consists of a series of platforms having suitable vertical drops along contour or on suitably graded lines across the general slope of the land. The vertical drop may vary from 60 cm to 180 cm depending upon the slope and soil conditions, as also economic width required for easy cultural operations. The material excavated from the upper part of the terrace is used in filling the lower part. A small 'shoulder' bund of about 30 cm height is also constructed along the outer ridge of the terrace. Bench terraces may be 'table top' or sloping outward or inward according to the rainfall of the tract is medium, poor or heavy and the soil and sub-soil are fairly absorptive or poorly permeable. Bench terraces should be adopted only on deep soil steeper than 15 per cent and in regions where land available for cultivation is scare or cash crop is grown. These terraces are generally very costly. Bench terracing is normally recommended on less than 33 per cent slope so that at least 2/3 of area is available for cultivation after loss of area in riser construction.

v) **Contour trenching:** It consists of series of 60 cm by 30 cm trenches excavated across the slope at suitable intervals. The excavated soil is deposited as a bund along the lower edge, and tree seedlings are planted along side the trench just above it for afforestation purpose.

vi) **Terrance outlet**: The terrace outlets are constructed for the safe disposal of excess runoff taking into account the maximum flood discharge of the catchment. Adequate outlets can usually be provided by protecting natural depressions with a cover of vegetation that will withstand the force of water moving down them. Accurate information regarding the maximum amount of runoff per unit of time, which may be expected in a given region is necessary for working out safe and economical plans for such outlets. To compute critical runoff from small catchment areas, the following rational formula proposed by Ramser may be used:

$$Q = C.I.A.$$

Where,

Q = Critical rate of runoff in cubic feet per second;

C = Runoff coefficient denoting the ratio of runoff to rate of rainfall

I = Rainfall intensity in cubic feet per second per acre, which happens to be almost equal to and may be represented by rate of rainfall in inches per hour; and

A = Area of watershed in acre. The runoff coefficient vary greately due to differences in slope, cover, soil material and state of cultivation.

vii) **Gully control**: About 2.3 m ha land in India is affected with severe gully erosion. Gullies are generally a secondary symptom, indicating misuse of the land over a long period of time causing functional disorder of the land. Gullies often start in the banks of natural watercourses that have been eroded to great depth. They extend back into the drainage area and grow deeper as they advance up the slope.

The first step in preventing the formation of gullies should include converting or permanent cover of such areas as are definitely too steep or eroded, the use of level or moderately sloping land for cultivated crops, and planting of natural depression or waterway to close growing perennial vegetation. The best known methods of controlling or preventing erosion are the vegetative controls, such as grass, mixture of grass and legumes and combination of vegetative and mechanical controls, such as contour cultivation, cover crops and crop rotations, and contour strip cropping, used alone or in combination with terracing where it is required, and the use of outlets for terrace-water disposal.

Although severely gullied land may have little immediate value, however the most economical and effective control of gullies and gullied areas can often be obtained by planting Kudzu (*Pueraria thumbergiana*) or tropical Kudzu (*Pueraria phaseoloides*) in good soil near the borders of the gullies so the vines will gradually encroach on the gullied areas.

Proper diversion trenches or perimeter bunds should also be constructed around the gully-head to prevent ingress of water into the gullies. In highly eroded areas, a master bund is erected right along the heads of ravines and gullies on the edge of cultivated fields to control the gully-head from extending into the field. This master bund is formed out of a graded channel suitably aligned on falling contour channel. There should be suitable outlets at the head of each minor gully so as to restrict water flow in the gully and thus, enable its being canalized.

Flumes of grass sod (sod flumes) may be successfully used to control over falls of limited head in gullies having relatively small drainage area. Sod checks are frequently used to stablize gully channels until the intervening areas become vegetated. In establishing vegetation in gullies the protection and assistance of temporary check dams, sometimes, required during the period of establishment. The materials such as brushwood, poles, wires, straw and stones usually used for constructing check dams. Permanent structures may be necessary in gullies that can not be adequately controlled by the vegetation. Several types of permanent structures like over flow dams, flumes and drop-outlets or soil saving dams may be used.

Acacia catechu, Wendlandia exerta, Saccharum spontaneum, Eriophorum comosum, Adhatoda vasica, and Woodfordia fruticosa have been found to establish themselves in dry, slightly xeric sites with very poor soil cover and debris of old mine material. In the lower reaches with good moisture regimes, mesophytic species, viz., *Leucaena leucocephala, Mallotus philippensis, Toona ciliata, Murraya koenigii, Adhatoda vasica and Thyanolena maxima* (broom grass) have established in their preferred habitats.

viii) **Reservoir or Ponds**: Farm ponds are the bodies of water made either by constructing an embankment across a water course or by excavating a pit or dug out.

*The runoff water is collected from the treated or untreated catchment and stored in a reservoir or farm pond. The stored water is utilized for supplemental irrigation during long dry spells at critical stages of plant growth. A supplemental irrigation of 5 + 2 cm given to a rainfed crop at reproductive stage can make all the differences between success or failure of the crop (Srivastava and Sharma, 1979; Faroda and Joshi, 1998).

The mechanical measure of **conserving ditches** (also known as **inverted contour bunding)** consists of trapezoidal ditches laid out on contours at the same interval as contour or graded bunds to serve as a terrace as well as storage structure in low rainfall deep soils.

The construction of reservoirs, ponds or storage tanks at suitable places to store water for water supply either for human and cattle consumption affords a practical method towards conservation of water in the catchment.

ix) **Stream bank protection:** Banks of streams or rivers get eroded and have to be canalised by easing the bank slopes or growing vegetation or by construction of stone or concrete pitching as necessary. Two types of structure can be erected for the protection of the bank: (i) structure that deflect the river current from the affected bank, thus aiding siltation. These are called 'spurs' or groynes, and (ii) structure that cover the affected surface and prevent further damage, like revetment and retaining wells, etc.

(C) Management of lands susceptible to wind erosion or already eroded by wind

Wind erosion is serious in areas where there is not enough vegetation to cover and protect the soil. This is the natural condition in arid lands and along the sand shores of lakes, oceans and rivers. Sand dunes are common landforms in such places. The basic principles of wind erosion control measures are: (a) to reduce wind velocity at soil surface, (b) to protect and conserve soil moisture, and (c) to trap the eroded soil particles. Soil and moisture conservation practices, like contour bunding, conservation, manures and fertilizers, mulching, shelter belts and agroforestry proved effective in providing more moisture for crop growth besides controlling erosion and soil loss (Faroda and Joshi, 1998). The following principles and practices should be adopted to avoid wind erosion.

i) **Conservation of moisture**: This can be done by applying organic matter, and ploughing of the lands during rains and mulching during the dry season. Dry farming should be practiced. The soil surface should be kept moist. One of the methods to overcome the problems of wind erosion could be the use of stubble mulch farming and growing wind strips.
ii) **Provision of soil cover**: Protection against wind erosion is best accomplished by the maintenance of a good vegetative cover on the soil during the dry seasons and seasons of high wind either by vegetative cover or mulching. Some sort of crops or vegetation should be allowed to remain on the surface of the soil. To break the impact of wind on the soil surface mulch

should be provided by leaving the residue of the crop on the land. Wind is more effective on smooth surface than on rough surface. Hence, if mulch is not possible, the surface should be left rough during the period of high winds which reduces the wind velocity. Ridges and furrows should be left as such during this period. Stubble aggregates large enough to resist the force of wind should be produced by application of organic matter, lime etc.

iii) **Provision of wind-breaks / shelter belts**: To reduce the effect of wind, strip cropping, ridges and shelter belts / wind breaks should be established. Strip- cropping and ridges are of temporary nature, whereas wind-breaks/shelter belts may be of permanent nature. Wind-breaks are effective up to 10 times their height. Grasses and shrubs should also be established for stablizing sand dunes. Tall plants species used as wind-breaks are *Dalbergia sissoo, Albizia lebbek, Acacia nilotica, Tamric spp. and Parkinsonia aculeata*. Decades of work done at Central Arid Zone Research Institute, Jodhpur showed that three-row shelter belts (*Cassia siamea - Albizia lebbeck - Cassia siamea*) put up across the wind direction were found to help in reducing wind speeds and checking soil erosion (Faroda and Joshi, 2000). Wind strips were found to help in controlling wind erosion. Comparing the different materials for wind strips, Mishra *et. al.*, (1960) reported that perennial grass sewan (*Lasiurus sindicus*), as wind strip increased the yield of pearlmillet by 9 per cent and also provided nutritional fodder for animals.

iv) **Stablization of drifting sand dunes**: The movement of sand dunes occur particularly during the dry rainless period with high wind velocity in arid-zones. In India about 22.7 per cent of total geographical area is occupied by sandy soils mainly in Rajasthan, Gujarat, Punjab and Haryana. Sand dunes are managed with protection against biotic interferences, mulch barrier in parallel strips, and planting suitable trees, grasses and shrubs to act as wind-breaks. The plant species selected for this purpose should have deeply burried rhizomes or prostrate growth with stem anchored by abundant adventitious fibrous roots for rehabilitation of sand dunes. The suitable trees are *Prosopis cineraria, P. chilensis, Acacia tortilis, Acacia radiana, Zizyphus mauritiana, Erianthus munja and Dalbergia sisso*. The bushes recommended for sand dunes are *Crotolaria burhia*, *Calligonum polygonoides*, *Aerva javanica* and *Ziziphus spp*. The suitable grasses are *Cenchrus cilliaris*, *Lasiurus sindicus*, *Panicum, turgidum* and *P. antidotale*.

v) **Artificial stabilizers**, like asphaltate clay, gravel picket fence, straw and iron netting may be used followed by permanent stablization by means of vegetation.

vi) **Organic manuring**: The same as in case of water erosion depending upon the degree of erosion produced.

The technique of afforestation for stabilizing sand-dunes had been developed by Central Arid Zone Research Institute (CAZRI) consists of followings:

- Protection against biotic interference.
- Treating the affected dunes by fixing micro-windbreaks in parallel strips or chess board design using locally available shrubs, starting from the crest down to the heel of the dunes. The 58 per cent area in western Rajasthan is covered by drifting or semi established sand dunes of up to a height of 100 m. These sand dunes are generally devoid of vegetation. The common locally available bushes are *Crotolaria burhia, Leptademia pyrotechnica* and *Ziziphus nummularia*.
- Afforestation of such dunes by direct seeding and planting. The direct seeding of grass species, like *Cenchrus ciliaris*, *Lasiurus sindicus* and *Saccharum munja* and creepers, like *Citrullus colosynthis* is carried out in lines 2 to 5 m apart, on the leeward side of micro windbreaks. The nursery raised seedlings are planted at the onset of monsoon.

The tree species suited for sand dune stabilization are *Acacia senegal*, *Acacia tortilis*, *Calligonum polygonoides*, *Colophospermum mopane*, *Callia glomerata* (syn. *Dichrostachys glomerata*), *Prosopis cineraria* and *Prosopis chibensis* (syn. *P. juliflora*)

Some other species suitable for semi-established sand dunes or established sand dunes are *Acacia salicina*, *Albizia lebbeck*, *Azadirachta indica*, *Cassia siamea, Eucalyptus camaldulensis*, *Hardwickia binata*, *Hloptelia integrifolia*, and *Tecomella undulata*.

REFERENCES

Anonymous. 1982. Soil Conservation Problem, Approach and Progress in India. Soil Conservation Division, Department of Agriculture and Cooperation, Ministry of Agriculture, Government of India, New Delhi.

Bennett, H. H., Bell, E.G. and Robinson, B.D. 1951. Raindrops and Erosion. Circular 895, United States Department of Agriculture, USA.

Bhushan, L. S. and Om Prakash. 1981. Soil loss and nutrient losses under different land use on 2% slope at Agra. Rep. C81 VCRIT, Dehradun, India, pp: 188.

Chepil, W.S. and Woodruff, N. P. 1963. The physics of wind erosion and its control. Advances in Agronomy, 15: 211-302.

CTIC. 1995. National Crop Residue Management Survey : Executive Summary. Conservation Tillage Information Center. West Lafayette, IN, pp: 21.

Dhruva Narayan, V. V. and Ram Babu. 1983. Estimation of soil erosion in India. Journal of Irrigation and Drainage Engineering, 109 (4): 419-34.

Dhruva Narayan, V. V. 1993. Soil and Water Conservation Research in India, ICAR, New Delhi, India.

Faroda, A. S. and Joshi, N. L. 1998. Sustaining Agriculture Through Resource Conservation In Arid Zones. First International Agronomy Congress - Proceedings, held at New Delhi on November 23-27, 1998 (*eds.* P. Singh, R. Prasad and IPS Ahlawat), ICAR, New Delhi, India, pp: 69-88.

Foster, G. R. and Meyer, L. D. 1972. A closed farm soil erosion equation for upland area. *In.* Sedimentation (*ed.* H. W. Shen) Eienstein Symposium. Ft Collins Co. 12: 1-19.

Hudson, N. 1971. Soil Conservation. B.T. Batsford Limited, London W1.

Mutchler, C. K. and Young, R.A. 1975. Soil detachment by rain drops. *In.* Present and Prospective Technology for Predicting Sediment Yields and Sources. ARS-S-40. USDA - Agriculture Resource Survey, pp: 117-27.

Lal, R., Wilson, G. F. and Okigbo, B. N. 1978. No - till farm after various grasses and leguminous cover crops in tropical alfisol. Journal of Crop Research, 1:71-84.

Lal, R., Wilson, G. F. and Okigbo, B. N. 1979. Change in properties of an alfisol by various cover crops. Soil science 127: 377-382.

Lal, R. and Kimble, J. M. 2000. Conservation Tillage: Prospects for Future - Invited Papers. *In.* International Conference on Managing Natural Resources for Sustainable Agricultural Production in the 21st Century held at New Delhi on February 14-18, 2000. pp: 116-125.

Middelton, H. E. 1930. Properties of soil which influence soil erosion. Technical Bulletin No. 175, USDA, Washington D.C., USA.

Mishra, M. N., Mishra, D. K. and Ram Prakash. 1966. Wind strip cropping increases grain yield in arid zone. Indian Farming, 17 (4): 39-40.

Ogrosky, H. O. and Mockus, V. 1957. The hydrology guide. Nat. Engng. Handbook, Section 4. SCS, US Department of Agriculture USA.

Rama Rao, M.S.V. 1962. Soil Conservation in India. ICAR. New Delhi, India.

Rama Rao, M.S.V. 1966. Watershed control of river valley projects. Journal of Soil and Water Conservation of India, 14: 1-5.

Sharda, V. N. and Jugal, G. P. 2006. Conservation Technologies for Sustaining Natural Resources. Handbook of Agriculture, ICAR, New Delhi, pp: 254-299.

Singh, G., Dayal, R. and Bhola, S. N. 1967. Soil and Water loss studies under different vegetative covers on 0.5 and 1.0% slopes at Kota. Journal of Soil and Water Conservation of India, 15 (3, 4): 17-23.

Singh, R. P. 1983. Farm Ponds. Bulletin No. 6. All India Coordinated Research Project for Dryland Agriculture, Hyderabad, India.

Srivastava, S. N. 1973. Peak rate of runoff from watershed of Soil Conservation Research Farm at Demotand, Hazaribagh. Journal of Soil and Water Conservation of India. 22-23: 28-31.

Srivastava, V. C. and Sharma S. S. 1979. Moisture use efficiency of barley of varying seeding rate at different nitrogen under rainfed condition. Andhra Agricultural Journal, 26 (3&4): 118-123.

Srivastava, V.C. and Mohsin, M.A. 1989. Effect of conventional and zero-tillage with and without mulch and weedicides on wheat and barley under rainfed upland situation. Journal of Research, BAU, 1(1): 9-14.

Smith, D. D. and Weischmeier, W. H. 1962. Rainfall erosion. Advances in Agronomy, 14: 109-48.

Stallings, J. H. 1957. Soil Conservation. Prentice Hall. Englewood Cliffs, New Jersey, USA.

Tejwani, K. G., Gupta, S. K. and Mathur, H. N. 1975. Soil and Water Conservation Research, ICAR, New Delhi, India.

Tiwari, K. M. 1964. Soil Conservation for Watershed protection and flood control. Journal of Soil and water conservation in India, 12: 25-31.

USDA. 1964. A Manual on Conservation for Soil and Water. Oxford Book Company, Calcutta / Kharagpur/New Delhi, India.

Vasudevaiah, R. D., Singh, S. P. and Gupta, D. P. 1965. Runoff- soil loss determination at Deochanda Experimental station. Journal of Soil and Water Conservation in India, 13: 34-46.

Venkateshwaradu, J. and Singh, S. N. 1982. A decade of Dryland Agricultural Research in India 1971-80. All India Coordinated Research Project for Dryland Agriculture, Hyderabad.

Vijay Lakshmi, K., Vittal, K. P. R. and Singh, R. P. 1982. A decade of dryland Agriculture Research in India 1971-80, AICRPDA, Hyderabad, India.

Wischmeier, W. H. and Smith, D. D., 1965. Predicting Rainfall Erosion Losses from Crop Land East of Rocky Mountains Guide for Selection of Practices for Soil and Water Conservation. Agriculture Handbook No. 282, USDA, USA.

Wischmeier, W. H. and Smith, D. D., 1978. Predicting rainfall erosion losses - A guide to Conservation Planning. Agriculture Handbook No. 537, USDA, USA.

Woodruff, N. P. and Siddoway, F. H. 1965. Wind erosion equation. Soil science Society of America Proceedings, 29: 602-608.

Young, R. A. and Weirsma, J. L. 1973. The role of rainfall impact on soil detachment and transport. Water Resource Research, 9: 1629-36.

CHAPTER 5

Waterlogged Soils

The term 'waterlogging' refers to a condition of short or long term water stagnation caused due to change in hydrologic regime, landscape, development activities, silting up of river beds, etc. Repeated flooding is yet another cause of waterlogging in coastal and flood-plain areas of major rivers. Increasing problems of short or long duration flooding over the years is attributed largely to deforestation in catchment areas, destruction of surface vegetation, changes in land use, increased urbanization and other development activities. The processes leading to flooding are being accentuated due to increased sedimentation and reduced capacity of drainage systems (Velayutham and Bhattacharya, 2000).

DISTRIBUTION OF WATERLOGGED SOILS IN INDIA.

In India, waterlogging of soils are more acute in northern Indo-Gangatic plains, mainly due to flat topography, frequent floods, canal irrigation and obstruction to natural drainage ways. Large areas get flooded during the monsoon season. There are also extensive low-lying areas in the rice growing coastal belts of eastern and southern India where poor drainage seriously affects crop production during the monsoon season. Accordingly to the Ministry of Agriculture (1976), an area of about 6 million ha are suffering due to the menace of waterlogging, excluding the areas affected by flooding and many localised areas along the coastal line due to periodic ingress of sea water during high tides. The soil resource mapping programe

(1:250,000 scale) completed by the National Bureau of Soil Survey and Land Use Planning, Nagpur indicate that the physical soil degradation in terms of waterlogging in each state has affected 5.7 million ha, representing 1.74 per cent of the total geographical area in the country. On the basis of extent of waterlogging soils in different states, Bihar tops the list (2000.8 th ha) followed by Andhra Pradesh (891.8 th ha), West Bengal (609.6 th ha) and Gujarat (523.6 th ha). Because of the flat nature of the country, large areas of north Bihar after the flood recede, water stagnates in a large number of depression. In West Bengal, districts like Midnapur, 24-parganas, Howrah, and Hoogly are affected. In Maharastra, large damaged area in the old Decaan canal exists due to waterlogging. Serious waterlogging problems exists in Chambal command areas of Rajasthan and Madhya Pradesh. Though southern Deccan Plateau Region is comparatively good in drainage, waterlogging is being experienced in low lying areas and canal irrigated areas. Some typical examples of areas affected by waterlogging are areas occupying lower topographic positions, areas along the rivers (*'Diara'* lands) and those lying beyond levees (*'Tal'* lands).

In ariad zones where irrigation has been practised for 6-80 years, the ground water table has risen by 17-25 m (Faroda and Joshi, 1998). The seepage of excess irrigation water on light textured soil having confining layer at shallow depth can easily cause the development of parched water bodies which are much above the regional ground water table and give rise to drainage and salinity problems. Such conditions exist in at a number of places of Indra Gandhi Canal at Rajasthan which irrigates the driest part of the country.

CAUSES OF WATERLOGGING

The main and associated anthropogenic and natural factors causing waterlogging are:

- transmission losses from main and branch canals and distributaries,
- faulty on-farm water management,
- lack of natural drainage (flat terrain on island basin with radial centripetal drainage pattern),
- upheaval in the river bed / bed of drainage channel,
- indiscriminate cultivation in bed of drainage channels and putting of fishing barriers,
- interception of natural drainage due to construction of roads and embankments,
- discharge of surplus canal water into channels with inadequate capacity or in inland basins lacking drainage outlet,
- non-utilization of ground water,
- inundations by marine and delta cycles, and
- inundations in coastal areas during cyclonic storms.

CHARACTERISTICS OF WATERLOGGED SOILS

The main features of the submerged soil is the formation of a thin oxidised layer of a few millimeters in thickness on the surface with a thick reduced layer beneath due to the presence of a layer of standing water. This brings about a marked effect on the physical, physico-chemical, chemical and biological properties of soil influencing growth and nutrition of plants, particularly rice.

Physico-chemical characteristics

i) **Reduction in gaseous exchange**: Waterlogging causes drastic curtailment of the normal process of gaseous exchange between soil and air resulting in a rapid disappearance of oxygen from the soil air and enrichment with carbon-dioxide, methane, hydrogen and nitrogen.

The oxygen content is relatively high in a layer extending from a few millimeters to one centimeter at the soil-water interface. Immediately beneath this oxygenated layer, the oxygen drops off sharply and with increasing depth it disappear completely. The oxygen rich surface layer i.e., 'oxidised layer' is characterised by a brown colour and the presence of typical oxidised ions, such as Fe^{+3}, Mn^{+4}, NO_3^-, SO_4^{--}, and the aerobic microorganisms. The oxygen deficient lower layer called 'reducing layer' is on the other hand, characterised by a dark grey colour and the presence of the reduced ions, viz., Fe^{+2}, Mn^{+2}, NH_4^+, $S^=$, etc., anaerobic or facultative anaerobic organisms.

Ammonium nitrogen form applied as fertilizer or released from organic matter is converted to nitrate in the oxygenated surface oxidising layer. The nitrate diffuses into the inorganic layer below it and is denitrified, thus, causing substantial losses of ammonium fertilizers in rice soils.

The excess concentration of carbon-dioxide in the waterlogged soil can influence the growth of rice plants through its direct effect on water absorption and nutrient uptake and indirectly through its solubilizing effect on nutrients.

ii) **Drop in redox potential**: Soil submergence brings about a sharp drop in redox potential with Eh values becoming negative. The drop in redox potential influences the availability of some of the nutrient elements, leads to the formation of some toxic substances and indirectly affect the absorption and assimilation of nutrients elements.

The state of reduction of a waterlogged soil can be defined quantitatively by measuring either the intensity factor or the capacity factor. The intensity factor is redox potential. The value of Eh is positive and high in a well drained soil, whereas it is low and often negative in a waterlogged soil.

The primary conditions required for soil reduction are absence of oxygen, presence of easily decomposable organic matter and presence and activity of anaerobic soil micro-organisms.

Ponnamperuma (1955) correlated the change in redox potential of soil on waterlogging with the soil properties as follows:

– Soils with high levels of NO_3, potential remains positive even several weeks after waterlogging.

- Soils with high level of manganese or low in organic matter, potential remains positive for a long time after submergence.
- Soils low in active ion and manganese and with high organic matter, potential becomes negative and reaches values of -0.2 to -0.3 volt within two weeks of waterlogging. Temperature above and below 25°C retards soil reduction which is most pronounced in acid soils.

iii) **Increase in electrical conductivity :** Solubilization of a number of ions on waterlogging increases the electrical conductivity of a soil due to the release of Mn^{+2}, Fe^{+2}, accumulation of NH^{+4}, HCO_3^- and organic acid anions and also due to the displacement of ions from the exchangeable complex by Fe^{+2} and Mn^{+2} and dissolution of $CaCO_3$ in calcareous soil by carbonic acid. Solubilization of ions increases in the early phase of submergence, reaches a maximum and then decreases to a fairly stable value dependant on the nature of the soil. The decline after the maximum is due to precipitation of Fe^{+2}, Mn^{+2}, either as hydroxides or carbonates, fall of partial pressure of CO_2 and the decomposition of organic acids.

iv) **pH**: There is increase in the pH of acid soils and decrease in that of alkali or calcareous soils under waterlogged condition. This has a great implication in nutrient availability as well as on the microbial activity in soil. Drainage and exposure to air reverse back the changes in waterlogged soils. Ponnamperuma (1966) related the increase in pH of acid soils to the intensity of reduction and the decrease in pH of alkali or calcareous soils to the partial pressure of CO_2.

Chemical characteristics

i) **Organic matter**: In waterlogged soils, decomposition of organic matter is carried out mainly by fucultative or obligate anaerobes, as the result of which rate of decomposition of organic matter and assimilation of carbon and other nutrients are much slower than in aerobic soils. The lower nitrogen requirement in anaerobic metabolism in waterlogged soil leads to a greater and more rapid release of ammonium nitrogen in soil solution, which might be one of the reason of good response of rice to added organic matter. Low nitrogen requirement in anaerobic respiration also means lower immobilization of fertilizer nitrogen.

ii) **Nitrogen**: In waterlogged soils due to lack of oxygen the process of transformation cannot proceed to the nitrate stage. In such soils, the process of mineralization is carried out mainly by faculative and obligate anaerobic bacteria. As they function at a lower energy level as compared to aerobic bacteria, both the rate of mineralization and the rate of immobilization is lowered and hence, there is a rapid accumulation of NH_3 in waterlogged soil. The waterlogged soil provides an environment highly suitable for denitrification to occur. This is one of the major mechanism by which nitrogen is lost from a waterlogged soil.

iii) **Phosphorus**: Patrick and Mahapatra (1968) has reviewed extensively the transformation and availability of phosphorus in waterlogged soil. It has been shown that waterlogging in a soil increases the concentration of water soluble phosphorus. In non-calcareous soils, this increase in water soluble phosphate is always associated with a lowering of redox potential and increase in Fe^{+2} concentration, which suggests a role of Fe^{+3} bound phosphate. In acid soils phosphate associated with Fe^{+3} and Al^{+3} predominate. The increase in concentration of the water soluble P, when acid soils are submerged, result from (a) hydrolysis of ferric and almunium phosphate owing to increase in pH on waterlogging, (b) reduction of Fe^{+3} to Fe^{+2} with the liberation of chemically bounded P and (c) release of P held by anion exchange on clays and hydrous oxides of iron and aluminium.

iv) **Iron and manganese**: According to Motomura (1969), important factors which influence the dynamics of iron and manganese in waterlogged soils are easily decomposable organic matter, amount of iron/manganese oxide, the redox potential, soil pH and soil temperature. The presence of nitrate retards the reduction of iron and manganese. It has been found that inspite of the tremendous reduction of iron and manganese to Fe^{+2} and Mn^{+2}, the concentration of Fe^{+2} and Mn^{+2} is soil solution in waterlogged soils is relatively small which may be due to the precipitation of these ions as hydroxides, carbonates or even sulphides. The increase in pH of acid soils on waterlogging favours the precipitation as hydroxides. According to Mandal (1961), the transformation of iron and manganese in waterlogged rice soils schematically represented the change in iron in waterlogged soils. He found that the insoluble Fe_2O_3 is first reduced to insoluble Fe^{+2} compounds, such as hydroxides and carbonates which in presence of sufficient CO_2 are transformed to bicarbonate which are soluble. Then the Fe^{+2} enters into the exchangeable complex and goes into soil solution.

v) **Other micronutrients**: The pH of an acid soil almost invariably increases to about neutrality after several days of waterlogging. The pH increase may lower the availability of Zn, B, Co but increase the availability of Mo. The uptake of Zn may also be lowered due to its interaction with P which is more soluble in waterlogged soil.

Microbial Activity

The waterlogged condition suppresses the bacterial and fungal population of the soil, but at the same time makes the environment suitable for the growth of algae, which are chlorophyll bearing photosynthetic organisms. Some species of blue green algae belonging to the family Cyanophyceae, which are of common occurance in rice fields are capable of fixing nitrogen from the atmosphere and they play an important role in the nitrogen economy of rice soils in tropical countries.

MANAGEMENT OF WATERLOGGED SOILS

A. Remedial measures

Remedial measures used are given below:

i) Drainage either surface or sub-surface or both to remove surplus water.
ii) Lining of canals to prevent seepage and rise of water-table.
iii) Sinking tubewell and utilizing water for irrigation, thereby lowering water table.
iv) Waterlogging lands contiguous to the sea-coast may be ameliorated for crop cultivation or used for fish culture.
v) Prevention of waterlogging and secondary salinization in the command areas of irrigation project.
vi) Need for erosion-control measures in the catchment areas.
vii) Judicious use of soil and water management in irrigated areas.

B. Production System

Submerged soils are primarily used all over the world for the production of rice crop, which is usually followed by a second and even a third crop of rice or an upland crop like wheat, maize, etc. (Mandal and Mandal, 2000). Deficiency of oxygen in the reducing layer does not create much disturbance to the normal functioning of the roots of rice because of their two types of adoption, viz., oxygen transport from the aerial parts to the roots and anaerobic respiration. Armstrong (1967, 1970) with the help of a mathematical model measured the oxygen flow across the roots of rice plants and found that it is sufficient to meet the oxygen requirement of root cells to oxidize the rhizosphere and protect the roots from the toxic effects of reduced substances.

Rice-cum-fish production system with its proper taking into account the compatibility of different varieties of rice with different species of fish may be useful for achieving sustainable agricultural production in waterlogged soils. Many waterlogged areas in Haryana are used for mixed farming, viz., rice and fish culture.

Special types of waterlogged conditions are seen in Malda Forest Division of West Bengal, where *beel* areas pose serious problem of afforestation. The lands here are fertile with Gangetic alluvial deposit, but most of the areas are low-lying, a good part of which except bottommost portions, remain completely dry for about eight months, whereas during the monsoon and soon after (i.e., mid July to mid-October) these areas remain totally submerged by the discharge of the rivers and accumulation of run-off from the surrounding catchment. These areas are locally known as *beel* and may be up to 5 Km wide and 10 km in length. The depth of water may be up to 5 m. Almost nothing grows except tall grasses. The first and most important operation is to stub out all tufts of tall grasses from the plantation area to reduce competition and to save the area from fire hazards during summer. Similar *beels* are common in the north-eastern states including Brahmputra valley of Assam and in the states of Manipur, Tirpura and Arunachal Pradesh. Planting of all plants in the waterlogged area is done on mounds of suitable diamensions so that they can escape complete submergence during rainy season. It is beneficial to apply fertilizers along with planting. In areas, where the water dries up after rains, it is observed that termites attack the seedlings of many species. Traditionally waterlogged areas are used for rice cultivation. In some areas which are subject to flooding in Assam, West Bengal and Bangaladesh the tall indigenous varieties of rice are still grown. These species have a unique feature of tallness and remain always above the water level.

REFERENCES

Armstrong, W. 1967. The relationship between oxidation-reduction potentials and oxygen diffusion levels in some water-logged organic soils. J. Soil Sci. 18(1): 27-34.

Armstrong, W. 1970. Rhizosphere oxidation in rice and other species: A mathematical model based on the oxygen flux component. Physiol. Plant 23 (3): 623-630.

Bremner, J. M. and Shaw, K. 1958. Deintrification in soil. II. Factors affecting deintrfication. J. Agric. Sci. 51:40-52.

De, P. K. and Sankar, S. N. 1936. Transformation of nitrate in waterlogged soil. Soil Sci. 42, 143-155.

Dennett, J. H. 1932. A preliminary note on the reaction of paddy soils. Malay. Agric. Jour. 20: 518.

Chakraborty, S. N. and Ghoshal, T. K. 1968. Anaerobic reduction and Phosphate solubility. J. Ind. Soc. Soil Sci. Vol. 16: 415.

Faroda, A. S. and Joshi, N. L. 1998. Sustaining agriculture through resource conservation and zones. *In* First International Agronomy Congress held on Nov. 23-27, 1998. at New Delhi, pp: 69-88.

Ghosal, T. K. and Chakraborty, S. N. 1967. Role of chelating agents in the release of phosphate from acid soils. J. Ind. Soc. Soil. Sci. Vol-15, 89.

Khan, S. K. and Mandal, L. N. 1973. Distribution of different inorganic forms of phosphorus in rice soil of West Bengal. J. Ind. Soc. Soil Sci. 21, 395-402.

Mandal, L. N. 1961. Transformation of iron and manganese in waterlogged rice soils. Soil Sci. 91, 121-126.

Mandal L. N. 1962. Levels of iron and manganese in soil solution and the growth of rice in waterlogged soils in relation to the oxygen status of soil solution. Soil Sci. 94, 387-391.

Mandal L. N. and Das, R. K. 1970. Transformation of applied water soluble phosphate in acidic low land rice soils. Soil Sci. 110, 263-267.

Mandal, L. N. and Nandi, S. N. 1971. Influence of intensity of soil reduction on the availability of iron, manganese and phosphorus and the growth and nutrition of rice. J. Ind. Soc. 19, 237-242.

Mandal L. N. and Mandal B. 2000. Management of submerged soils for sustainable agricultural production. Extended summaries Vol 1: Invited Papers International Conference on Managing Natural Resources for sustainable Agricultural Production in the 21st Century held on February 14-18, 2000, New Delhi, India.

Ministry of Agriculture and Irrigation, 1976. Report Part III National Commission on Agriculture, Government of India, New Delhi, India.

Subrahmanyan, V. 1927. Biochemistry of waterlogged soils. I. The effect of waterlogging on the different forms of nitrogen on the reaction, on gaseous relationship and on bacterial flora. J. Agric. Res. 17, 429-467.

Motomerra, S. 1969. Dynamic behaviour of ferrous iron in paddy soil. J.A.R.Q. 4,12-17.

Patric, W.H. (Jr.) 1964. Extractable iron and phosphorus in submerged soil at controlled redox potentials. Proc. 8th. Inter. Congr. Soil Sci. Bucharest, 4:605-610.

Patric, W.H. (Jr.) and Mahapatra, I.C. 1968. Transformation and availability to rice of nitrogen and phosphorus in waterlogged soils. Adv. Agro. 20, 323-359.

Ponnamperuma, F. N. 1955. The chemistry of waterlogged rice soils in relation to the growth and nutrition of rice. Ph.D. thesis, Cornell Univ.

Ponnamperuma, F. N. 1966. Influence of redox potential and partial pressure of carbon dioxide on pH values and the suspension effect of flooded soils. Soil Sci. 101, 421.

Ponnamperuma, F. N. 1966. Ionic strengths of the solutions of flooded soils and other natural acqueous solutions from specific conductance. Soil Sci. 102, 408.

Schollenberger, C. J. 1928. Manganese as an active base in the soil. Soil Sci 25, 357-358.

Sinha, N. P., and Mandal, S. C. 1955. Different types of phosphates in some Bihar Soils, Jour. & Proc. Inst. Chem. (India), Vol. XXVII, 105.

Singh, R. N., Sinha, H. & Sahi, B. P. 1968. Fractionation of soil inorganic phosphorus. J. Ind. Soc. Soil Sci., Vol. 16: 371.

Turner, F. T. & Patric, W.H. (Jr.) 1968. Chemical changes in water-logged soils as a result of oxygen depletion. Trans. Int. Congr. Soil Sci. 9th Vol. 4:53-65.

Velayutham, M. and Bhattacharya, T. 2000. Soil Resource Management. In Natural Resource Management for Agricultural Production in India (eds. J.S.P. Yadav and G. B. Singh). International Conference on Managing Natural Resources for sustainable Agricultural Production in the 21st Century held on Feb. 14-18, 2000 at New Delhi, India, pp: 3-135.

CHAPTER 6

Peatlands or Organic Soils

Peatlands or organic soils (Histosols) are currently recognised as fragile ecosystems. The utilization of these lands results in a decline or loss in productivity with a significant impact on entire ecosystem. Technically agriculture is possible in such soils. However, the high capital input required is an economic deterrant (Padmanabhan et al., 2000)

Most of the soils of the world contain some amount of organic matter, but there are certain soil which contain a huge amount of organic matter and are called organic soils. On the basis of organic matter, an arbitrary boundary has been created between organic and mineral soils. Soils containing up to 20 per cent organic matter as the higher limit are known as mineral soils, whereas those containing above 20 per cent and less than 50 per cent not so well decomposed organic matter are known as organic soils. Soils with well decomposed organic matter are known as muck.

DISTRIBUTION

Globally, wetlands occupy about 18.96 million sq. km of which 2 per cent made up of organic soils. In Asia 220, 189 sq. km of land surface comprise pit or histosols. India accounts for 0.27 per cent of its land area (Eswaran, 1986 ; Soil Survey staff 1999).

DEVELOPMENT

These soils develop in marshes, bogs and swamps by anaerobic decomposition of plants growing there. These water-bodies grow a number of plants like mosses, ferns, grasses, shrubs or even trees. On death, these plants accumulate at the bottom of water-bodies and undergo slow anaerobic decomposition forming peat soils. In a fairly deep bog or lake, all the above mentioned plants do not grow at the same time but grow in order of succession. First, there are mosses, algae, and floating water plant. When the lake is filled to some extent by their debries and become shallow, ferns, grasses, shrubs and trees invade them in order of succession.

CLASSIFICATION

Peats are classified on the basis of plant source from which they are formed. As these plants differ greatly in their make up, histology and chemical composition, so does the peats formed by them.

i) **Sedimentary peat**: This is formed mainly from the faeces of aquatic animals feeding on mosses, algae and floating flora. Sedimentary peats are rubbery in nature, very fine in texture, greenish in colour when freshly dug which turn deep black on exposure.
ii) **Fibrous peat**: This is formed when the lake has become shallower enough to allow the growth of ferns, grasses, seeds and shrubs. The decomposition product of these plants are brown to black in colour, powdery and may contain dicomposed fibrous tissues of the above mentioned plants.
iii) **Woody peat**: When the marshes (low lying wetland) become shallower enough to allow the growth of aquatic trees, the peat is formed by the decomposition of leaves, branches, trunks of these trees and their undergrowth. Apart from containing the powdery and fibrous organic matter, the woody peat also contains decomposed to nearly fresh woody parts of the aquatic trees.

iv) **Sphagnum moss peat**: It is kind of fibrous peat formed by the anaerobic decomposition of sphagnum moss, a vegetation of low fertility status.

The layering may be in order of sdementary peat, fibrous peat, woody peat, from bottom upwards of the lakes.

According to Eswaran (1986), peat or histosols in the tropics are found in two geomorphic positions:

a) **Topogenous peats**: These are formed at high elevations where the temperatures are low and extreme humid conditions prevail. Rates of accumulation exceed decomposition of organic matter in these environments and peats formed even on slope as steep as 40 per cent. Most of the extensive peat lands are present in the swamps of the tropics.
b) **Ombrogenous peats**: These peats are formed in basin-shaped topographic conditions favouring rainwater accumulation. In this environment, decomposition of organic matter is retarded by the oxygen free conditions of the permanent water table. Most ombrogenic peats occur along the coast, with frequent inundation with seawater. With growth of the peat domes, fresh water replaces the barkish conditions.

CHARACTERISTICS

(a) Physical

1) **Colour**: It varies from dark brown to deep black. Freshly exposed sedementary peats are green in colour.
2) **Bulk density**: The parent material for peat is either wood or soft tissues, such as the roots of reeds, leaves and twigs. Consequently, there are differences in the bulk densities of the sapric material developed from wood (0.4 - 0.8 g per cubic cm) and that derived from softer tissues (<0.3 g per cubic cm) as compared to 1.3 to 1.5 of mineral soils.

3) **Waterholding capacity**: It has got high water holding capacity (200 to 400%) as compared to mineral soils. The sedementary peat has maximum water holding capacity and woody peat the least.
4) **Structure**: These soils have very low plasticity as compared to mineral soils. Hence, they form small aggregates and become crumb when ploughed.
5) **Shrinkage on drainage and cultivation**: On drainage and cultivation, they begin to oxidise quickly and the shrinkage in depth due to organic matter decomposition may be from 1 to 3 inch per year depending upon the climate of the region.

(b) Physico-chemical

1) A characteristic feature of many tropical peat is its acidic nature and consequently, low available nutrient content status. In contrast to temperate peats derived from moss or sedges, tropical peats are woody making mechanical cultivation extremely difficult.
2) The partial decomposed wood fragments are interspersed with humified organic materials with high water holding and cation exchange capacities.
3) The zero point of net charge is very low and so anion exchange is insignificant. However, the large surface area and ability to hold cations imply that heavy metals are retained by the system.

(c) Chemical

1) **Ash content**: It may vary from less than 5 per cent to 40 per cent. Sphagnum moss has generally less than 5 per cent ash content.
2) **Organic matter**: It may vary from more than 20 per cent to 95 per cent.
3) **C/N ratio**: The C/N ratio of peat soils are high as compared to mineral soils. The minimum C/N ratio found is 20.

4) **Total N**: It varies from 1 to 3.5 per cent. Sphagnum moss contains as low as 0.5 per cent.

MANAGEMENT OF PEATLANDS

The major constraints to sustainable development of peatlands include short range variability in soil depth.

Considering the constraints imposed upon by the physical and chemical nature of peatlands attaing sustainable production in such areas require tremendous investment and technical input. High fertilizer inputs are required in addition to careful management of water table (Padmanabhan *et al.,* 2000). Despite the fragility of the system, peat swamp forest is rapidly being cleared and drained for other land-uses, especially agriculture (Prentice and Parrish, 1992).

In Asia, land-use on the peat lands is low-input subsistence - based systems. Many communities and cities have developed around swamps and rivers, e.g. Kuala Lampur, Bangkok, Jakarta, Mumbai, Shanghai, etc. with rice as a staple food. Lowlands have always in demand and in recent decade and with much of the lowland already utilized, people are forced to look for land on steep slopes. Large areas of peats in Malaysia and Indonesia have been or are being cleared for plantation of rubber and oil palm.

Subsistence based agricultural system would not be able to provide the level of input required for sustainable management of peatlands. However, commercial based agricultural systems appear to thrive in areas dominated by shallow peat. Probably the most sustainable use of peatlands is managed forestry. Timber planting must be done in a planned manner and replanting should be a mandatory follow-up process.

The coastal peatlands of the tropics protected on the seaward side by raised ridges, which are standard beach deposits. The inland swamp are fed with fresh water from the rivers which are admixed with sea water. The adjoining sea and rivers provide adequate supply of fish and sea-food. Peat swamp forests are widespread in southeast Asia, mostly occuring on flat coastal plains, although it also extends up river valleys and occurs locally in the highlands. Prentice and Parrish (1992) showed the importance of peat swamp forests as a habitat for wildlife, support large mammals, atleast seven primate species, corcodile and rich bird fanua that differs from that of dry forest types.

REFERENCES

Anderson, J.A.R. 1974. The structure and development of the peat swamp of Sarawak and Brunei Journal of Tropical Geography, 18: 7-16.

Driessen, P. M. Soepraptohardjo, M. and Pons, L. J. 1975. Formation, properties, reclaimation and agricultural use in Indonesian ombrogenous lowland peats. International Peat Symposium, Tel Aviv, Israel.

Eswaran, H. 1986. Classification and management of Histosols. *In*. Eswaran *et al.* (Eds). Proceedings of 2nd International Symposium on Peatland soil. Thailand/Malaysia. Public Department of Land Development, Thailand, pp:15-26.

Eswaran, H. 1992. Role of soil information in meeting the challenges of sustainable land management. Journal of Indian Society of soil Science, 40: 6-24.

Eswaran, H. and Reich. 1998. Global Land Resource Database. In preparation. USDA National Resources Conservation Service.

Fuchmans, C. H. 1986. The peat-water problem. *In.* Peat and Water - Aspects of Water Retention and Dewatering in Peat *ed.* C.H. Fuchmans. Elsevier Applied Science Publication, London, New York, pp: 331-360.

Hwai, Y. B., Padmanabhan, E and Crabtree, J. M. 1999. Role of subsurface terric materials on composition and behaviour of Histosols (in review).

Padmanabhan, E., Eswaran, H. and Reich, P. 2000. Conservation Issues of Managing peatlands in Asia. *In.* International Conference on Managing Natural Resources for sustainable Agricultural Production in the 21st Century held on February 14-18, 2000, New Delhi, India, pp: 155-163.

Parkyn, L., Stoneman, R. E. and Ingram, A. 1998 (eds.). Conserving Peatlands. Publication Commonwealth of Agriculture Bureau Inc. U. K., pp:50.

Plucknet, D. L. and Winkelmann, D. L. 1995. Technology for sustainable agriculture. Scientific American 150: 182-186.

Plyusnin, I. I. 1970's. Reclaimative Soil Science. Foreign Lang. Publishing House, Moscow.

Prentice, C. and Parish, D. 1992. Conservation of peat swamp forest: a fortotten ecosystem. Malayan Nature Journal (Malaysia), 45: 128-144.

Soil Survey Staff. 1999. Soil Taxonomy. A system of Soil classification for making and Interpreting Soil Surveys. Natural Resource Conservation Service. US Dept. of Agriculture., Handbook 336, Second ed. pp: 89, US Government Print Office, Washington D.C., USA.

Subject Index